SYSTEM DESIGN AUTOMATION

T0205638

System Design Automation

Fundamentals, Principles, Methods, Examples

Edited by

Renate Merker

and

Wolfgang Schwarz

Technische Universität Dresden, Germany

KLUWER ACADEMIC PUBLISHERS
BOSTON / DORDRECHT / LONDON

A C.I.P. Catalogue record for this book is available from the Library of Congress.

ISBN 978-1-4419-4886-1

Published by Kluwer Academic Publishers,
P.O. Box 17, 3300 AA Dordrecht, The Netherlands.

Sold and distributed in North, Central and South America
by Kluwer Academic Publishers,
101 Philip Drive, Norwell, MA 02061, U.S.A.

In all other countries, sold and distributed
by Kluwer Academic Publishers,
P.O. Box 322, 3300 AH Dordrecht, The Netherlands.

Printed under founding of the Deutsche Forschungsgemeinschaft.

Printed on durable acid-free paper

Preface

Modern microelectronic technology provides the possibility to implement complex systems on single chips. Generally such systems include analogue and digital hardware, embedded software, peripheral controllers, sensors and actuators. Due to their complexity they cannot be analyzed and designed without the aid of computers. With the rapid progress of technology, the development of new principles, dedicated methods and tools for computer-aided design has become an indispensable and permanent requirement. As a consequence, the challenge facing the design community consists in the developing of methodologies and appropriate tool-support techniques for a systematic and efficient design automation.

Design automation of electronic and hybrid systems is accordingly a steadily growing field of interest and a permanent challenge for researchers in Electronics, Computer Engineering and Computer Science.

This book presents some recent results in design automation of different types of electronic and mechatronic systems.
It deals with various topics of design automation, ranging from high level digital system synthesis, through analogue and heterogeneous system analysis and design, up to system modeling and simulation. Design automation is treated from the aspects of its theoretical fundamentals, its basic approach and its methods and tools. Several application cases are presented in detail.

The book is organized as follows:

In the first chapter **High-Level System Synthesis** (Digital Hardware/Software Systems, Application Cases) embedded systems, distributed systems and processor arrays as well as hardware-software codesign are treated. Three special application cases are also discussed in detail.

The second chapter **Analog and Heterogeneous System Design** (System Approach and Methodology) faces issues of the analysis and design of hybrid systems composed of analog and digital, electronic and mechanical components.

In chapter three: **System Simulation and Evaluation** (Methods and Tools) object-oriented modeling, analog system simulation, including fault-simulation, parameter optimization and system validation are considered.

The contents of the book is based on material presented at the Workshop System Design Automation (SDA 2000) organized by the Sonderforschungsbereich 358 of the Deutsche Forschungsgemeinschaft at TU Dresden.

The editors are grateful to all authors for their excellent cooperation. Dirk Fimmel and Jan Müller deserve our special gratitude for the editorial production of this book.

Dresden, December 2000   Renate Merker,   Wolfgang Schwarz

Contents

SYSTEM SIMULATION AND EVALUATION

Methods and Tools

HIGH-LEVEL SYSTEM SYNTHESIS

Digital Hardware/Software Systems

Synthesis and Optimization of Digital Hardware/Software Systems

Jürgen Teich, Datentechnik, Universität Paderborn, 33095 Paderborn, Germany
teich@date.upb.de

Abstract

In this introductory paper to the field, it is our goal to provide a new unified look at synthesis problems that is independent from the level of abstraction like system, RTL, and logic (for refinements targeted to hardware), or process- and basic block level (for refinements targeted to software). For each level of our model called "double roof", synthesis requires the solution of three basic problems, namely *allocation* (of resources), *binding*, and *scheduling*. Based on the "double roof" model, we present a graph-based formulation of the tasks of system-synthesis: Contrary to former approaches that consider system-synthesis as a bi-partition problem (e.g., earlier hardware/software partitioning algorithms), we consider also as well the allocation of components like micro- and hardware coprocessors as part of the optimization problem as scheduling of tasks including communication scheduling. The approach is flexible enough to be applied to different other abstraction levels. Finally, we introduce the problem of design space exploration as a new challenge in synthesis. For the typically multi-objective nature of synthesis problems, not only one optimum is wanted, but an exploration of a complete front of optimal solutions called Pareto points.

1 Introduction

Technology roadmaps for the design of integrated circuits foresee systems including more than a dozen of different processor cores and dedicated co-processors, so-called systems on a chip (SoC), to be feasible to fit on a single die within the next few years. In order to be able to design such complex systems correctly and thus exploit these remarkable technological advances, design (CAD) tools must also be able to specify, refine, verify, and synthesize such coarse-granular components correctly including their interplay.

Here, the gap between technology and tool maturity and acceptance is increasing instead of decreasing. Although placement and routing tools as well as logic synthesis tools have become popular and accepted within most industrial design flows, the above mentioned complexity requires even higher abstraction levels to be automized, in particular the so called system-level. A research area that has recently recognized and defined the new problems encountered when designing complete systems comprising as well software programmable microprocessors as dedicated hardware components (e.g., an MPEG4 video-compression system) is called *hardware/software codesign*. Although systems including hardware and software have been developed in common since a long time, hardware and software have been often developed separately from each other, resulting in either under- or overdesigned systems.

Obviously, the proporties of different alternative solutions must be explored early in the design phase in order to avoid later inconsistencies or complex redesign phases. In order to achieve this, *system-level synthesis* tools and *design space exploration* define new challenges for new design tools.

R. Merker and W. Schwarz (eds.), System Design Automation, 3-26.
© 2001 *Kluwer Academic Publishers.*

2 The double roof model

In order to cope with increasing design complexity of hardware systems, many models to categorize and identify different hierarchical design problems like, e.g., in Gajski's famous Y-chart model [8], have been elaborated.

Here, we introduce a synthesis-oriented model called *double roof*, where one side of the roof represents those abstraction layers that are typically encountered in hardware design, whereas the other side represents those layers that are typical to software synthesis for embedded systems (see Fig. 1).

Also, unlike Gajski's Y-chart [8] that distinguishes three different views of hardware objects in the design of hardware systems, namely behavior, structure, and geometry, our model [20] focuses only on the first two views that are important to understand synthesis problems.

The double roof model, as shown in Fig. 1, looks like a roof with two layers. The top layer represents the *behavioral layer*, whereas the lower layer represents the *structural layer*.

There are many points on the roof each of which may be identified with an *abstraction level*. Before introducing typical abstraction levels encountered during synthesis of hardware as well as of software during the design of an embedded system, we shortly summarize our view of synthesis problems that is independent from the abstraction layer considered.

An informal definition of *synthesis* may be given as follows: *Synthesis is the refinement of a behavioral specification into a structural specification at a certain abstraction level. The main synthesis tasks are independent from the level of abstraction and may be classified as [20]*

- allocation *of resources,*

- binding *behavioral objects to allocated* structural objects, *and*

- scheduling *of* behavioral objects *on the resources they are bound to. A schedule may be a function that specifies the absolute or relative time interval, a behavioral object is executed, or just an order relation for the execution of several objects (complete order or partial order, priorities, etc.).*

Mathematically, allocation is a problem of selection, and binding and scheduling are typically assignment problems (e.g., to resources, time steps, respectively), possibly subject to complex constraints.

In the following, we identify those abstraction layers which are typical for designing embedded systems.

2.1 Abstraction levels in system design

The design of a complex hardware/software system requires the introduction of different abstraction levels that allow the sucessive *refinement* of a given specification into manageable pieces.

In Fig. 1, we have tried to identify not all, but typical abstraction levels encountered during the synthesis of an embedded hardware/software system. These will be explained in the following:

- *system*: Models on the system-level describe the complete system to be designed on the level of networks of communicating subsystems (e.g., processors, ASICs, dedicated hardware units, memories, buses, etc.), each realizing a part of the behavioral system specification (e.g., algorithms, tasks).

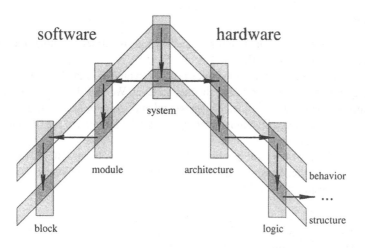

Figure 1: Double roof model: abtraction levels and views during the design of embedded hardware/software systems

- *architecture*: The architecture level belongs to the hardware side of the roof. Models at this level describe communicating functional blocks (often called RT (*register transfer* level blocks) that implement coarse granular arithmetic and logical functions.

- *logic*: This level also belongs to the hardware domain. Here, models describe netlists of logic gates and registers that implement Boolean functions and finite state machines.

- *module*: This abstraction level belongs to the software side of the roof. Models at this level describe the interaction of functions with complex behavior, e.g., process networks [16], task-level graphs [19], languages with support for threads, e.g., Java, etc. that are mapped to either a uni-processor with or withour support of a multi-tasking operating system, or a multi-processor architecture in software.

- *block*: The block-level may also be associated with the software domain. Behavioral models at this level typically include programs, functions, procedures as encountered in high-level languages that are refined to the instruction-level of the target-processor on which the code is to be executed.

Although one may distinguish many more different abstraction levels beyond those distinguished in Fig. 1, the design of a complex embedded system may be described by a process that consists of a sequence of *refinement* steps each of which is a translation of a behavioral description at a certain abstraction level into a structural description of the same level. Such a refinement basically adds structural information about the implementation. Using this information, the behavior may be sucessively refined further on the next lower level of abstraction.

Typically, a system is neither developed *top-down*, nor *bottom-up*. Also, some components may already exist at lower-levels, and must or should be used in the system for some reasons. Or, a team engineering a system cooperatively may simultaneously work on different abstraction levels so that at a certain point of time, not all system components have reached the same amount of refinement.

2.2 Synthesis

Instead of performing a refinement step by hand, synthesis allows to refine a system under design semi- or even fully automatically.

In the introduction, we claimed that the major tasks of synthesis, namely, allocation, binding, and scheduling, are independent from the abstraction level. We will show that the major differencies are only the object granularities of *behavioral* and *structural objects*, the *optimization algorithms* used for deriving allocations, bindings, and scheduling, and the *objectives* to optimize in each synthesis phase.

Table 1 gives a summary of typical granularities of behavioral objects (BO), structural objects (SO), and what the synthesis task at the level system (S), module (M), block (B), architecture (A), and logic (L), respectively, is known as.

	S	M	B	A	L
Behavioral Objects	process networks	tasks, threads	procedure, function	$+,*,\geq,\leq$ $,=,\cdots$	$\vee,\wedge$
Structural Objects	processors, ASICs, buses, memories,	uni-processor program	assembler	datapath + controller	gate netlist
task	hw/sw-partitio-ning	process scheduler	HLL-compiler	high-level synthesis	logic synthesis

Table 1: Characteristics of synthesis objects and synthesis tasks at different levels of abstraction

Here, we would like to identify at least for certain levels what the tasks of synthesis are, namely for the levels system (see also Section 3), module, and architecture in Fig. 1.

2.2.1 System-Level

Fig. 2 shows a system on a chip (SoC) design that integrates a processor core cell (left), a gate-array (right) for realization of logic, memory (ROM), and some periphery blocks (e.g., A/D-Converter, Timer, etc.) on a single chip.

At the system-level, it is an important task to decide which and how much of the active chip area should be used and how, e.g., how much area should be reserved for hardware (gate-array), how much for processors and which processor core(s) should be used at all. Such a decision of the system-level architectural components is the allocation step at the system-level.

Behavioral objects at this level are typically computation intensive tasks as may be specified using a design language such as Java, SystemC, process languages [11], or domain-specific languages such as ESTEREL [3] for reactive systems [9, 2], or SDL [18] for telecommunication systems. Recently, also graphical formalisms based on block-oriented data-flow networks [15], [16], StateCharts [10], and many others have been proposed to specify the behavior of a system under design at the system-level.

2.2.2 Module-Level

At the module-level, a number of tasks or program threads has to be refined and mapped to a single or multiprocessor software implementation that may have been allocated at the system-

Figure 2: System on a chip (SoC) design example of a hardware/software system. Courtesy: Texas Instruments, cDSP

level. Here, allocation of memory segments for program and data is important, especially for above mentioned single-chip solutions where memory must be integrated on chip, hence is expensive.

In [21, 25], we study the effect of code optimizations from data flow graph specifications by inlining or subroutine calls as well as the effect of loop nesting and context switching on a single processor target that is used as a main component in a memory and cost-critical hardware/software design, e.g., a single-chip solution. The methodology begins with a given *synchronous dataflow graph* [14] as used in many rapid prototyping environments as input for code generators for programmable digital signal processors (PDSPs) [5, 13, 17].

Example 2.1 *A practical example is a sample-rate conversion system. In Fig. 3, a digital audio tape (DAT), operating at a sample rate of 48 kHz is interfaced to a compact disk (CD) player operating at a sampling rate of 44.1 kHz, e.g., for recording purposes, see [23] for details on multistage sample rate conversion.*

$$\boxed{A}\xrightarrow{1\quad 1}B\xrightarrow{2\quad 3}C\xrightarrow{2\quad 7}D\xrightarrow{8\quad 7}E\xrightarrow{5\quad 1}\boxed{F}$$
CD   DAT

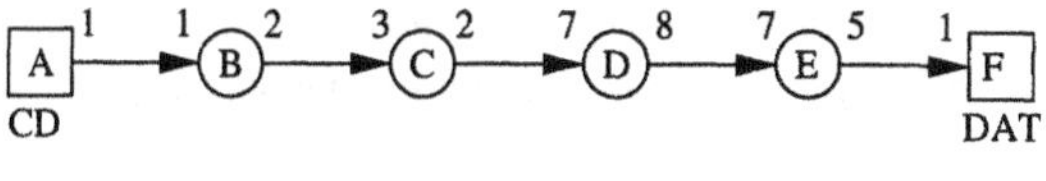

Figure 3: CD to DAT conversion process (SDF) graph. The numbers associated with the arcs denote the number of tokens produced by the predecessor, consumed by the sucessor actor, respectively, when firing.

Here, a *schedule* is a sequence of actor firings. A properly-constructed SDF graph is compiled by first constructing a finite schedule S that fires each actor at least once, does not deadlock, and produces no net change in the number of queue tokens associated with each arc. When such a schedule is repeated infinitely, we call the resulting infinite sequence of actor firings a *valid periodic schedule*, or simply *valid schedule*.

Example 2.2 *For the CD to DAT graph in Fig. 3, the minimal number of actor firings for obtaining a periodic schedule is $q(A) = q(B) = 147$, $q(C) = 98$, $q(D) = 28$, $q(E) = 32$, $q(F) = 160$. The schedule $(\infty(7(7(3AB)(2C))(4D))(32E(5F)))$ represents a valid schedule.*

Each parenthesized term $(n\ S_1\ S_2\ \cdots\ S_k)$ is referred to as *schedule loop* having *iteration count* n and *iterands* $S_1, S_2, \cdots, S_k$. We say that a schedule for an SDF graph is a *looped schedule* if it contains zero or more schedule loops. A schedule is called *single appearance schedule*, or simply *SAS* in the following, if it contains only one appearance of each actor.

Example 2.3 *The schedule $(\infty(147A)(147B)(98C)(28D)\ (32E)(160F))$ is a valid SAS for the graph shown in Fig. 3.*

From a given schedule, code may be synthesized automatically as follows: For each actor in a valid schedule S, we insert a code block that is obtained from a library of predefined actors or a simple subroutine call of the corresponding subroutine, and the resulting sequence of code blocks (and subroutine calls) is encapsulated within an infinite loop to generate a software implementation. Each schedule loop thereby is translated into a loop in the target code. If a code block for an actor does not exist already in a library of the target processor, it has to be synthesized at the block-level (see Fig.1) by a target-specific compiler.

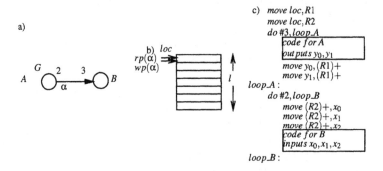

Figure 4: SDF graph a), memory model for arc buffer b), and Motorola DSP56k-like assembly code realizing the schedule $S = (\infty(3A)(2B))$

Example 2.4 *For the simple SDF graph in Fig. 4a), a buffer model for realizing the data buffer on the arc α as well as a pseudo assembly code notation (similar to the Motorola DSP56k assembly language) for the complete code for the schedule $S = (\infty(3A)(2B))$ is shown in Fig. 4b), c) respectively. There is a location loc that is the address of the first memory cell that implements the buffer and one read ($rp(\alpha)$) and write pointer ($wp(\alpha)$) to store the actual read (write) location.*

The notation do #N LABEL denotes a statement that specifies N successive executions of the block of code between the do-statement and the instruction at location LABEL. First, the read pointer $rp(\alpha)$ to the buffer is loaded into register R1 and the write pointer $wp(\alpha)$ is loaded into R2. During the execution of the code, the new pointer locations are obtained without overhead using autoincrement modulo addressing $((R1)+, (R2)+)$. For the above schedule, the contents of the registers (or pointers) is shown in Fig. 5.

In Section 4, we will continue the above example and show what algorithms may be used for optimizing the memory and execution time requirements at the module-level.

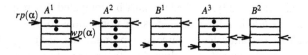

Figure 5: Memory accesses for the schedule $S = (\infty A(2AB))$

2.2.3 Architecture-Level

Example 2.5 *The following example is a behavioral specification of a hardware design for solving the differential equation* $y'' + 3xy' + 3y = 0$ *in the interval* $[x_0, a]$ *with step size dx and initial values* $y(x_0) = y$, $y'(x_0) = u$ *numerically.*

The specification in VHDL could look as follows: (x_0 corresponds to port signal x_in*,* $y(x_0)$ *to* y_in*,* $y'(x_0)$ *to* u_in*,* dx *to* dx_in *and* a *to* a_in*):*

```
ENTITY dgl IS
  PORT (x_in, y_in, u_in, dx_in, a_in: IN REAL;
        activate: IN BIT;
        y_out: OUT REAL);
END dgl;

ARCHITECTURE behavioral OF dgl IS BEGIN
  PROCESS (activate)
    VARIABLE x, y, u, dx, a, x1, u1, y1: REAL;
  BEGIN
    x := x_in; y := y_in; u := u_in;
    dx := dx_in; a := a_in;
    LOOP
      x1 := x + dx;
      u1 := u - (3 * x * u * dx) - (3 * y * dx);
      y1 := y + (u * dx);
      x := x1; u := u1; y:= y1;
      EXIT WHEN x1 > a;
    END LOOP;
    y_out <= y;
  END PROCESS;
END behavioral;
```

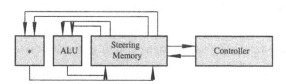

Figure 6: Example of a structural view at the architecture level

Fig. 6 gives an example of the structure that may be obtained as a result of architecture (high-level) synthesis, The datapath contains a multiplier, an ALU (arithmetic-logical unit), register memory, as well as a unit for connecting registers to functional units (e.g., bus- or multiplexer-based).

First, resources must be allocated (i.e., functional units (FUs) like multiplier units and ALUs). Then, operations of the specification must be bound to the allocated resources (operations to

FUs, variables to registers). Finally, a schedule must be computed that respects the resource constraints. Here, a *schedule* is a function that assigns operations to time steps. In Fig.7, an example of a schedule for two allocated resources of type multiplier, and one ALU is shown. For scheduling, any resource-constrained scheduling algorithm ranging from a simple list scheduler to ILP (integer linear programming) techniques may be used.

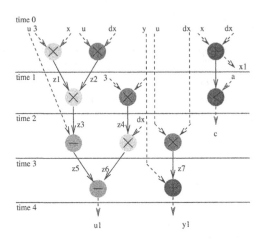

Figure 7: Data dependence graph for the specification in Example 2.5 and schedule with latency $L = 4$

3 System-synthesis

We use the term system-synthesis for denoting synthesis at the system-level. Previous approaches to system-synthesis considered synthesis subproblems like the binding of coarse-granular tasks at the system-level to either hardware or software (so-called hardware/software partitioning). Here, bottom-up solutions including clustering as top-down approaches including greedy strategies have been proposed. In [20], [4], we proposed a formal graph-theoretic approach to the generalized synthesis problem the main concepts of which are summarized next. A synthesis specification model consists of three main components:

- The problem that should be mapped onto an architecture as well as the class of possible architectures are described by means of a universal *dependence graph* $G(V,E)$.

- The user-defined mapping constraints between tasks and architectures are specified in a specification graph $G_S(V_S, E_S)$. Additional parameters which are used for formulating the objective functions and further functional constraints are annotated to the components of G_S.

- Associated with nodes and edges of the specification graph are activations which characterize the allocation and binding.

Definition 3.1 (Dependence Graph) *A dependence graph is a directed graph $G(V,E)$. V is a finite set of nodes and $E \subseteq (V \times V)$ is a set of directed edges.*

For example, the dependence graph to model the data flow dependencies of a given specification will be termed *problem graph* $G_P = (V_P, E_P)$ [4]. Here, V_P contains nodes which model either functional operations or communication operations. The edges in E_P model dependence relations, i.e. define a partial ordering among the operations.

Example 3.1 *One can think of a problem graph as the graph obtained from a data flow graph by inserting communication nodes into some edges of the data flow graph (see Fig. 8). These nodes will be drawn shaded in the examples.*

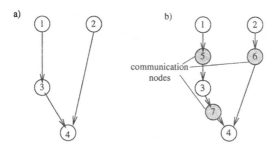

Figure 8: A data flow graph a) and the corresponding problem graph b)

The problem graph may be either acyclic or cyclic, e.g., an iterative problem graph with weights denoting initial data tokens on edges, etc. Also, different semantics may be associated with nodes, e.g., control flow nodes (evaluating data-dependent conditions and activating alternative sucessor nodes depending on such conditions) and data flow nodes (activating by default all sucessor nodes) may be distinguished. Thus, variations of control/data flow graphs also belong to the class of problem graphs that may be considered.

Now, the architecture including functional resources and buses can also be modeled by a dependence graph termed *architecture graph* $G_A = (V_A, E_A)$. V_A may consist of two subsets containing functional resources (hardware units like adder, multiplier, RISC processor, dedicated processor, ASIC) and communication resources (resources that handle the communication like shared buses or point-to-point connections). An edge $e \in E_A$ models a directed link between resources. All the resources are viewed as *potentially allocatable* components.

Example 3.2 *Fig. 9a) shows an example of an architecture consisting of three functional resources (RISC, hardware modules HWM1 and HWM2) and two bus resources (one shared bus and one unidirectional point-to-point bus). Fig. 9b) shows the corresponding architecture graph G_A.*

The *specification graph* will also be used to define *binding* and *allocation* formally.

Definition 3.2 (Specification Graph) *A specification graph is a graph $G_S = (V_S, E_S)$ that consists of 2 dependence graphs $G_i(V_i, E_i)$ for $1 \leq i \leq 2$ and a set of mapping edges E_M. In particular, $V_S = \bigcup_{i=1}^{2} V_i$, $E_S = \bigcup_{i=1}^{2} E_i \cup E_M$ where $E_M \subseteq V_1 \times V_2$.*

Consequently, the specification graph consists of two dependence graphs and mapping edges from nodes of the first graph G_1 to nodes of the second graph G_2. We associate the problem graph G_P with G_1 ($G_1 = G_P$) and the architecture graph G_A with G_2 ($G_2 = G_A$). The edges of E_M represent user-defined mapping constraints in the form of a relation: 'can be implemented by'.

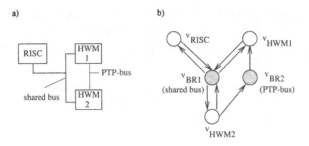

Figure 9: An example of an architecture a), and the corresponding architecture graph G_A b)

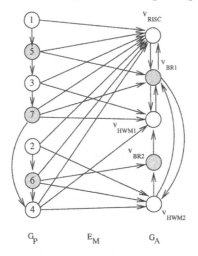

Figure 10: An example of a specification graph G_S

Example 3.3 *Fig. 10 shows an example of a specification graph using the problem graph of Example 3.1 (left), and the architecture graph of Example 3.2 (right). The edges between the two subgraphs represent the mapping edges E_M that describe all possible node bindings. For example, operation v_1 can be executed only on v_{RISC}. Operation v_2 can be executed on v_{RISC} or v_{HWM2}. Also, communication v_7 can be executed by v_{BR1} or within v_{RISC} or v_{HWM1}.*

This way, the model of a specification graph allows a flexible expression of the expert knowledge about useful architectures and mappings.

In order to describe a concrete mapping, i.e., an *implementation*, the term *activation* of nodes and edges of a specification graph is defined. Based on this definition, *allocation*, *binding* and *scheduling* will be formally defined in the next section.

Definition 3.3 (Activation) *The activation of a specification graph $G_S(V_S, E_S)$ is a function $a : V_S \cup E_S \mapsto \{0, 1\}$ that assigns to each edge $e \in E_S$ and to each node $v \in V_S$ the value 1 (activated) or 0 (not activated).*

The activation of a node or edge of a dependence graph describes its use. In the examples used so far, all nodes and edges of the problem graph contained in G_S may be necessary, i.e., activated. The determination of an implementation can be seen as the task of assigning activity values to each node and each edge of the architecture graph. An activated mapping edge represents the fact that the source node is implemented on the target node.

Now, the term implementation will be formally defined as well as the main tasks of synthesis, namely *allocation*, *binding*, and *scheduling*.

Definition 3.4 (Allocation) *An allocation α of a specification graph is the subset of all activated nodes and edges of the dependence graphs, i.e.,*

$$\alpha = \alpha_V \cup \alpha_E$$

$$\alpha_V = \{v \in V_S | a(v) = 1\}$$

$$\alpha_E = \bigcup_{i=1}^{2} \{e \in E_i | a(e) = 1\}$$

Definition 3.5 (Binding) *A binding β is the subset of all activated mapping edges, i.e.,*

$$\beta = \{e \in E_M | a(e) = 1\}$$

Definition 3.6 (Feasible Binding) *Given a specification G_S and an allocation α. A* feasible binding β *is a binding that satisfies*

1. *Each activated edge $e \in \beta$ starts and ends at an activated node, i.e.,*

$$\forall e = (v, \tilde{v}) \in \beta \; : \; v, \tilde{v} \in \alpha$$

2. *For each activated node $v \in \alpha_V$ with $v \in V_1$, exactly one outgoing edge $e \in E_M$ is activated, i.e.,*

$$| \{e \in \beta \, | e = (v, \tilde{v}), \tilde{v} \in V_2\} \, | = 1$$

3. *For each activated edge $e = (v_i, v_j) \in \alpha_E$ with $e \in E_1$,*

   - *either both operations are mapped onto the same node, i.e.,*

$$\tilde{v}_i = \tilde{v}_j \quad \text{with} \quad (v_i, \tilde{v}_i), (v_j, \tilde{v}_j) \in \beta$$

   - *or there exists an activated edge $\tilde{e} = (\tilde{v}_i, \tilde{v}_j) \in \alpha_E$ with $\tilde{e} \in E_2$ to handle the communication associated with edge e, i.e.*

$$(\tilde{v}_i, \tilde{v}_j) \in \alpha_E \quad \text{with} \quad (v_i, \tilde{v}_i), (v_j, \tilde{v}_j) \in \beta$$

It is useful to determine the set of feasible allocations and feasible bindings in order to restrict the combinatorial search space.

Definition 3.7 (Feasible Allocation) *A* feasible allocation α *is an allocation that allows at least one feasible binding β.*

Theorem 3.1 (Feasible Binding [20, 4]) *The determination of a feasible binding is NP-complete.*

The proof may be found in [20, 4].

Finally, it is necessary to define a *schedule*. Let $delay(v, \beta)$ denote the (estimated) execution time of the operation associated to node v of the problem graph G_P ($=G_1$). In order to be as general as possible at this point, we suppose that the execution time depends on a particular binding β. In other words, the execution time of an operation depends on the resource where it is going to be executed.

Definition 3.8 (Schedule) *Given a specification G_S containing a simple, acyclic data flow graph as problem graph $G_1 = G_P$, a feasible binding β, and a function delay which determines the execution time $delay(v, \beta) \in \mathbb{Z}^+$ of a node $v \in V_P$. A schedule is a function $\tau : V_P \mapsto \mathbb{Z}^+$ that satisfies for all edges $e = (v_i, v_j) \in E_P$:*

$$\tau(v_j) \geq \tau(v_i) + delay(v_i, \beta)$$

$\tau(v_i)$ may be interpreted as the start time of the operation of node $v_i \in V_P$. For example, the execution time of a communication node denotes the number of time units necessary to transfer the associated data on the bus resource it is bound to. Usually, these values depend not only on the amount of data transferred but also on the capacity of the resource, for example the bus width and the bus transfer rate. Therefore, the delay may depend on the actual binding.

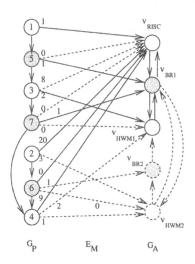

Figure 11: An example of an implementation of the specification given in Fig. 10

Example 3.4 *Consider the case that the delay values of a node $v_i \in V_P$ only depend on the binding of that particular node. Then the delay values can be associated with the edges E_M (see Fig. 11). The execution times of all operations on different resources are shown. For example, operation v_3 takes 8 time units if executed on the RISC (v_{RISC}) but 2 time units if mapped to the hardware module HWM1. Note that internal communications (like the mapping of v_5 to v_{RISC}) are modeled to take zero time.*

Definition 3.9 (Implementation) *Given a specification graph G_S, a (valid) implementation is a triple (α, β, τ) where α is an allocation, β is a feasible binding, and τ is a schedule.*

Example 3.5 *Fig. 11 shows an implementation of the specification depicted in Fig. 10. The nodes and edges not allocated are shown dotted, as well as the edges $e \in E_M$ that are not activated. The allocation of nodes is $\alpha_V = \{v_{RISC}, v_{HWM1}, v_{BR1}\}$ and the binding is $\beta = \{(v_1, v_{RISC}), (v_2, v_{RISC}), (v_3, v_{RISC}), (v_4, v_{HWM1}), (v_5, v_{RISC}), (v_6, v_{BR1}), (v_7, v_{BR1})\}$.*
Note that communication modeled by v_6 can be handled by the functional resource v_{RISC} as both predecessor node (v_2) and successor node (v_4) are mapped to resource v_{RISC}. A schedule is $\tau(v_1) = 0$, $\tau(v_2) = 1$, $\tau(v_3) = 2$, $\tau(v_4) = 21$, $\tau(v_5) = 1$, $\tau(v_6) = 21$, $\tau(v_7) = 4$.

With the model introduced previously, the task of system synthesis can be formulated as an optimization problem.

Definition 3.10 (System Synthesis) *The task of* system synthesis *is the following optimization problem:*

minimize $f(\alpha, \beta, \tau)$,
subject to

$\quad$ α *is a feasible allocation,*
$\quad$ β *is a feasible binding,*
$\quad$ τ *is a schedule, and*
$\quad$ $g_i(\alpha, \beta, \tau) \geq 0$, $i \in \{1, \ldots, q\}$.

The constraints on α, β and τ define the set of valid implementations. Additionally, there are functions $g_i, i = 1, \ldots, q$, that together with the *objective (or cost) function f* describe the optimization goal.[1]

Example 3.6 *The specification graph G_S may consist only of a problem graph G_P and an architecture graph G_A. Consider the task of latency minimization under resource constraints, i.e., an implementation is searched for that is as fast as possible but does not exceed a certain cost MAXCOST. To this end, a function cost : $V_A \mapsto \mathbb{Z}^+$ is given which describes the cost $(\tilde{v})$ that arises if resource $\tilde{v} \in V_A$ is realized, i.e., $\tilde{v} \in \alpha$. The limit in costs is expressed in a constraint $g_1(\alpha, \beta, \tau) = MAXCOST - \sum_{\tilde{v} \in \alpha} cost(\tilde{v})$. The corresponding objective function may be $f(\alpha, \beta, \tau) = max\{\tau(v) + delay(v, \beta)) \mid v \in V_P\}$.*

The objective function may be arbitrary complex and reflect the specific optimization goal. Likewise, the additional constraints g_i can be used to reduce the number of potential implementations. In [20, 4], many examples and extensions of the above model are given. For example, it is shown how to express sharing of resources, hierarchy, etc.

4 Design Space Exploration

Because of the accepted use of synthesis tools on lower design levels of abstraction, exploration becomes the next step in order to prevent under- or overdesigned systems. Typically, a

---

[1]We will see in Section 4 that the cost function is often a vector function (multiobjective optimization).

system has to obey many constraints and should optimize many different design objectives and constraints simultaneously such as execution time, cost and area, energy consumption, weight, etc.

A single solution that optimizes all objectives simultaneously is very unlikely to exist. Instead, it should be possible to first explore different optimal solutions or approximations thereof, and subsequently select and refine one solution out of these.

As will be shown by examples from different levels of abstraction, design space exploration is important on any of the abstraction levels of the double roof model.

In this area, multiobjective optimization problems have to be solved. An approach that uses Evolutionary Algorithms (EAs) [7, 1] for design exploration of Pareto-optimal fronts can be found in [4] for system-level synthesis (a generalization of hardware/software partitioning), or for optimal code synthesis for DSP processors from data flow graphs [21], [24], [25], [26].

Unfortunately, existing tools are either too specialized to be used for different synthesis problems, too tightly coupled to tools that are used for evaluating the quality of a design point, heavily dependent on architecture assumptions, design abstraction, etc. such that a reuse of such tools is simply impossible.

Here, we present the structure of a versatile tool for design space exploration called EXPLORA [6] that may be easily incorporated into any level of design abstraction. The flexibility results by addressing the following problems and requirements:

- Formal (functional) quantification of the nature of design space exploration processes involving synthesis tasks.

- Clear separation between:

  - Problem-specific parameters (e.g., dimension of exploration space, metrics, cost function, etc.)
  - Independence of synthesis algorithm and implementation language
  - Independence of optimization (exploration) algorithm and implementation language
  - Visualization support

- Finally, it should be easy to couple such a design space exploration tool to existing environments.

In the following, we give a characterization of design space exploration processes using synthesis tools. In Section 4.2, we present the mathematical background to formalize the process of generic design space exploration, and introduce the structure of EXPLORA. Finally, we present one example of exploration on the architecture-level and one example on the module-level.

4.1 Characterization of design space exploration processes

Two of the basic requirements of a flexible tool for design space exploration are a) the exchangeability of the optimization algorithm that is used for exploration of the design space and b) its adaptability to different synthesis tools that may be used to compute the quality of points in the design space concerning cost, speed, and other metrics. A natural distinction is to split the process of design space exploration into three main tasks: The first one contains all synthesis tool specific behavior, the second concerns the optimization algorithm for evaluating solutions and selecting new design points in the design space with the goal to obtain a high diversity (covering) of optimal points. The third module manages the exploration process itself using handles to the other modules.

Since the optimization algorithm needs a cost function which rates a given result produced by the synthesis tool, it is useful to split the module with the optimization algorithm into a second module for computing the cost function (see Fig. 12). This way, the cost function can easily be changed by the user, too, without the need to exchange the optimization algorithm.

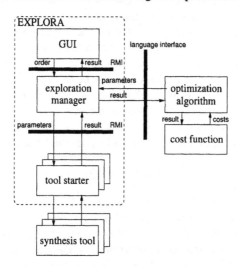

Figure 12: Structure of a program for design space exploration

Figure 12 shows the structure of a tool for generic design space exploration. The exploration manager starts a synthesis tool with certain parameters and obtains the synthesis results of this tool. This result is then forwarded to the optimizer that is used during the exploration. This algorithm provides the next parameter set(s) (new design point(s)) to explore. The optimization module in turn uses a cost function which rates a given result. Finally, it is also desirable to have a graphical user interface (GUI) that gathers and visualizes the progress and results during the exploration process.

Figure 12 shows the three different data structures `parameters`, `result` and `costs`. The `parameters`-object characterizes a design point and can be described by a set of parameters needed by the synthesis tool to perform a synthesis. Each parameter set characterizing a different design is also called a *design point*.

The exchanged data-structures may be explained as follows: A `result`-object represents that part of the synthesis tools output which is important to control the exploration flow and which is needed by the user to rate this output. Basically, it consists of a set of quantities representing the properties of the synthesis tool output. The `costs`-object may often be simply described by a tuple of real or integer numbers which must be properly interpreted by the optimization algorithm to rate a certain `result`.

4.2 Functional description of the exploration process

4.2.1 Tool starter module abstraction

Each synthesis tool needs some input files to perform a synthesis and produces some output files as a result. During the exploration, possibly not the whole input should be modified (e.g., the design specification stays the same) and not all parts of the output (e.g., command log files) should be taken into consideration. So the task of the tool starter in Fig. 12 is to produce the complete input needed by the synthesis tool given a parameter set and to extract the desired result quantities from the output returned by the synthesis tool after its completion.

The tool starter sits on top of the corresponding synthesis tool and behaves like an independent tool to its invoker, so its behavior can be described by a function as follows:

Let a certain synthesis tool starter have n parameters $p_1, ..., p_n$ with the domain P_i for the parameter p_i, $i = 1, ..., n$. Hence, a design point may also be characterized by an n-tuple $\vec{p}$ without loss of generality. Let the corresponding tool produce m result quantities $q_1, ..., q_m$ that may be represented by a tuple $\vec{q}$. Let Q_i be the domain of the quantity q_i, $i = 1, ..., m$. If there are no other constraints, then $P = P_1 \times P_2 \times ... \times P_n$ is called the *design space* $Q = Q_1 \times Q_2 \times ... \times Q_m$ the result space. The behavior of this synthesis tool starter can be described by a function

$$synth : P \to Q \tag{1}$$

Example 4.1 *Let the number of allocated functional units (FUs) of each type be the parameter of an architecture (high-level) synthesis tool which, given these parameters, tries to construct a schedule with minimal latency that satisfies the resource constraints as imposed by the allocation. Thus, the synthesis tool returns the latency and the area requirement of the resulting circuit. Let two types of FUs be available, namely a multiplier (parameter p_1) and an ALU (parameter p_2). Suppose the name of the VHDL file is a string that is a third parameter. Hence, a design point $\vec{p}$ is a three-dimensional tuple here. The domain of the function synth would be*

$$P = P_1 \times P_2 \times \{file\_name.vhdl\}$$

with $P_1, P_2 \subseteq N$, and the range of result values $Q = R^+ \times R^+$. For each valid VHDL file and for each value of $p_1 \in P_1, p_2 \in P_2$, the function synth would produce a pair of values for area and latency, denoted by $\vec{q} = (area(\vec{p}), latency(\vec{p}))$.

4.2.2 Cost function abstraction

For rating a synthesis result, the optimization algorithm needs a cost function which builds a tuple of costs from a tuple of result quantities.

Let $c_1, ..., c_l$ be l cost quantities. Let C_i be the value range of cost parameter c_i, $i = 1, ..., l$. The range of values of the resulting cost tuple $\vec{c}$ is C, $C = C_1 \times C_2 \times ... \times C_l$. The cost function is then

$$cost : Q \to C \tag{2}$$

Example 4.2 *Continuing the previous example, one cost function would be the weighted one dimensional function ($l = 1, m = 2$)*

$$cost(area(\vec{p}), latency(\vec{p})) = 0.7 * area(\vec{p}) + 0.3 * latency(\vec{p}).$$

This way, the cost function would weight the area of a design point $\vec{p}$ more than its latency time and thereby force the exploration in a direction which rather would produce smaller than faster designs.

Example 4.3 *Let us consider, without loss of generality, a multi-objective minimization problem with m result parameters for each design point $\vec{p}$ of dimension n and l objectives. Let $\vec{q} = synth(\vec{p})$.*

$$\text{Minimize} \quad \vec{c} = cost(\vec{q}) = (cost_1(\vec{q}), \dots, cost_l(\vec{q})) \tag{3}$$

where $\vec{q} = (q_1, \dots, q_m) \in Q$ and $\vec{c} = (c_1, \dots, c_l) \in C$ are tuples with $c_i = cost_i(\vec{q})$, $i = 1, \cdots, l$. $\vec{a} \in Q$ is said to dominate $\vec{b} \in Q$ (also written as $\vec{a} \succ \vec{b}$) iff

$$\forall i \in \{1, \dots, l\}: \quad cost_i(\vec{a}) \le cost_i(\vec{b}) \quad \wedge$$
$$\exists j \in \{1, \dots, l\}: \quad cost_j(\vec{a}) < cost_j(\vec{b}) \tag{4}$$

$\vec{a}$ covers $\vec{b}$ ($\vec{a} \succeq \vec{b}$) iff $\vec{a} \succ \vec{b}$ or $cost(\vec{a}) = cost(\vec{b})$. All design points $\vec{p}_i \in P$ with the property that $q_i = synth(\vec{p}_i)$ is not dominated by any other $q_j = synth(\vec{p}_j)$, $\vec{p}_j \in P$, are called nondominated. Pareto-optimal *points (or simply Pareto points) are the nondominated design points of the entire search space P.*

For design space exploration, a useful cost function is to let $cost(\vec{x})$ be equal to the number of design points explored so far that dominate $\vec{x}$. Hence, after the exploration, all explored points with cost zero are (approximations of) Pareto-optimal points.

4.2.3 Optimization module abstraction

The task of the optimization module is to produce a set of new parameter sets (design points) to explore next given a set of design points and their cost.

If a given optimization algorithm needs v different synthesis results to generate w new design points, then the behavior of the optimization module could be described by a function

$$opt : (P \times C)^v \to P^w \tag{5}$$

Example 4.4 *Consider an optimization (exploration) algorithm that is population-based, e.g., a variant of an Evolutionary Algorithm that simply selects the best result from n given result objects using a certain cost function and produces n identical copies of this optimal design point as offspring, however, with random variations in its parameters. In the next iteration of the exploration, this set of mutated design points would be used by the tool starter and produce new synthesis results. This way, the algorithm would implement a function*

$$opt : (P \times C)^n \to P^n$$

with $C = cost(synth(P))$.

Example 4.5 *A well-known local search technique for solving hard combinatorial problems is simulated annealing [12]. Here, the algorithm decides based on a single result object which new design point in the neighborhood will be investigated next: $opt : (P \times C)^1 \to P^1$. Hence, the exploration describes simply a path in the design space.*

4.2.4 Exploration manager

The exploration manager has just administrative tasks. It doesn't need to know about the specific implementations of the other modules, it just knows about their interfaces which must be common to all implementations of these modules.

To perform the exploration using a given synthesis tool, the exploration manager has to properly invoke the corresponding functions.

As exploration is an iterative process, the obvious idea is to have a loop somewhere in the manager module that rates previously generated results using the cost function (*cost*), starts the optimization module with these ratings and the corresponding parameter sets (*opt*), and calls *synth* for each new design point to be explored.

Conceptually, this means the successive execution of the functions *cost*, *opt* and *synth* in each iteration. Generally, this isn't straightforward, since in the case if *opt* needs v different (cost,parameter) tuples to produce w new parameter sets, the function *cost* must be invoked v times, and after that *synth* must be executed w times.

Additionally, the values v and w are generally not constant, and the process of the exploration must be take care of that, see [6] for details.

4.3 Example: Exploration at architecture-level

During architecture synthesis, a tradeoff has to be made, typically between speed (latency) and cost. For designs where the cost may be estimated well by the number of allocated functional units, the quality of different design points (objective space) is shown in Fig. 13 for the case study introduced in Example 2.5.

Let us assume that the latency of each resource is one cycle for each operation and that the delay in the steering and control logic is negligible. Assume furthermore, that a multiplier requires 5 units of area and an ALU 1 unit, controller, memory, steering logic one additional unit.

In Fig. 13, different synthesis solutions obtained for different allocations are shown, corresponding to allocations of $p_1 = 1$ multiplier, $p_2 = 1$ ALU (design point $(1,1)$), with cost and area of 7, point $(1,2)$ with latency 7 and area 8, point $(2,1)$ with latency 5 and area 12, and point $(2,2)$ with latency 4 and area 13. Hence, the parameter space is $P = \{1,2\} \times \{1,2\}$. For this small space, a simple enumeration algorithm is chosen in the optimizer of EXPLORA to obtain the synthesis results for all design points in this small parameter space.

In Fig. 13, the point $(1,2)$ is not a Pareto point because it is dominated by point $(1,1)$ which has equal latency, however smaller area requirements. All other points are Pareto points.

4.4 Example: Exploration at module-level

Continuing our code optimization methodology, we describe the application of an Evolutionary Algorithm for automatic design space exploration of uni-processor DSP implementations of SDF graphs. They turn out to be ideal candidates for discrete solution spaces with complex constraints and multi-objective cost functions. Here, the (three-dimensional) cost vector is more complicated and includes the three objectives progam memory requirements, data memory requirements, and execution time overhead.

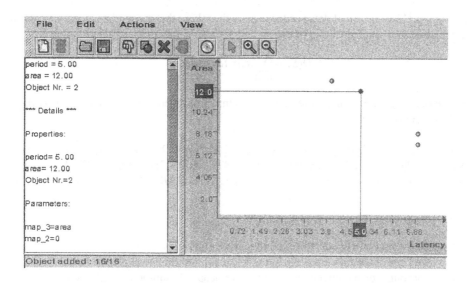

Figure 13: Example of design space exploration using EXPLORA at the architecture-level

4.4.1 Program memory overhead $P(S)$

Assume that each actor N_i in the library has a program memory requirement of $w(N_i) \in N$ memory words. Let $flag(N_i) \in \{0,1\}$ denote the fact whether in a schedule, a subroutine call is instantiated for all actor invocations of the schedule ($flag(N_i) = 0$) or whether the actor code is inlined into the final program text for each occurence of N_i in the code ($flag(N_i) = 1$). Hence, given a schedule S, the program memory overhead $P(S)$ will be accounted for by the following equation:[2]

$$
\begin{aligned}
P(S) \quad = \quad & \sum_{i=1}^{|V|} (app(N_i, S) \cdot w(N_i) \cdot flag(N_i)) \\
+ \quad & (w(N_i) + app(N_i, S) * P_S) \cdot (1 - flag(N_i)) \\
+ \quad & P_L(S)
\end{aligned}
\tag{6}
$$

In case one subroutine is instantiated ($flag(N_i) = 0$), the second term is non-zero adding the fixed program memory size of the module to the cost and the subroutine call overhead P_S (code for call, context save and restore, and return commands). In the other case, the program memory of this actor is counted as many times as it appears in the schedule S (inlining model). The additive term $P_L(S) \in N$ denotes the program overhead for looped schedules. It accounts for a) the additional program memory needed for loop initialization, and b) loop counter incrementation, loop exit testing and branching instructions. This overhead is processor-specific, and in our computations proportional to the number of loops in the schedules.

4.4.2 Buffer memory overhead $D(S)$

We account for overhead due to data buffering for the communication of actors. The simplest model for buffering is to assume that a distinct segment of memory is allocated for each arc of a

---

[2] $app(N_i, S)$: number of times, N_i appears in the schedule string S.

given graph.[3] The amount of data needed to store the tokens that accumulate on each arc during the evolution of a schedule S is given as:

$$D(S) = \sum_{\alpha \in A} max\_tokens(\alpha, S) \tag{7}$$

Here, $max\_tokens(\alpha, S)$ denotes the maximum number of tokens that accumulate on arc α during the execution of schedule S.

Example 4.6 *Consider the schedule in Example 2.3 of the CD to DAT benchmark. This schedule has a buffer memory requirement of* $1471 + 1472 + 982 + 288 + 325 = 1021$*. Similarly, the buffer memory requirement of the looped schedule* $(\infty(7(7(3AB)(2C))(4D))(32E(5F)))$ *is 264.*

4.4.3   Execution Time Overhead $T(S)$

With execution time, we denote the duration of execution of one iteration of a SDF graph comprising $q(N_i)$ activations of each actor N_i in clock cycles of the target processor.[4]

Here, we account for the effects of (1) loop overhead, (2) subroutine call overhead, and (3) buffer (data) communication overhead in our characterization of a schedule. Our computation of the execution time overhead of a given schedule S therefore consists of the following additive components:

Subroutine call overhead: For each instance of an actor N_i with $flag(N_i) = 0$, we add a processor specific latency time $L(N_i) \in N$ to the execution time. This number accounts for the number of cycles needed for storing the necessary amount of context prior to calling the subprogram (e.g., compute and save incremented return address), and to restore the old context prior to returning from the subroutine (sometimes a simple branch).[5] *Communication time overhead*: Due to static scheduling, the execution time of an actor may be assumed fixed (no interrupts, no I/O-waiting) necessary), however, the time needed to communicate data (read and write) depends in general a) on the processor capabilities, e.g., some processors are capable of managing pointer operations to *modulo buffers* in parallel with other computations.[6], and b) on the chosen buffer model (e.g., contiguous versus non-contiguous buffer memory allocation). In a first approximation, we define a penalty for the read and write execution cycles that is proportional to the number of data read (written) during the execution of a schedule S. For example, such a penalty may be of the form

$$IO(S) = 2 \sum_{\alpha = (N_i, N_j) \in A} q(N_i) produced(N_i) T_{io} \tag{8}$$

---

[3]In [22], we introduced different models for buffer sharing, and efficient algorithms to compute buffer sharing. Due to space requirements, and for matters of comparing our approach with other techniques, we use the above simple model here.

[4]Note that this measure is equivalent to the inverse of the throughput rate in case it is assumed that the outermost loop repeats forever.

[5]Note that the exact overhead may depend also on the register allocation and buffer strategy. Furthermore, we assume that no nesting of subroutine calls is allowed. Also, recursive subroutines are not created and hence disallowed. Under these conditions, the context switching overhead will be approximated by a constant $L(N_i)$ for each module N_i or even to be a processor-specific constant T_S, if no information on the compiler is available. Then, T_S may by chosen as an average estimate or by the worst-case estimate (e.g., all processor registers must be saved and restored upon a subroutine invocation).

[6]Note that this overhead is then highly dependent on the register allocation strategy.

where T_{io} denotes the number of clock cycles that are needed between reading (writing) 2 successive input (output) tokens.

Loop overhead: For looped schedules, there is in general the overhead of initializing and updating a loop counter, and of checking the loop exit condition, and of branching, respectively. The loop overhead for one iteration of a simple schedule loop L (no inner loops contained in L) is assumed a constant $T_L \in N$ of processor cycles, and its initialization overhead $T_L^{init} \in N$. Let $x(L) \in N$ denote the number of loop iterations of loop L, then the loop execution overhead is given by $O(L) = T_L^{init} + x(L) \cdot T_L$. For nested loops, the total overhead of an innermost loop is given as above, whereas for an outer loop L, the total loop overhead is recursively defined as

$$O(L) = T_L^{init} + x(L) \cdot \left(T_L + \sum_{L' \text{ evoked in} L} O(L') \right) \tag{9}$$

The *total loop overhead* $O(S)$ of a looped schedule S is the sum of the loop overheads of the outermost loops.

Example 4.7 *Consider the schedule* $(\infty(3(3A)(4B)) (4(3C)(2D)))$, *and assume that the overhead for one loop iteration* $T_L = 2$ *cycles in our machine model, the initialization overhead being* $T_L^{init} = 1$. *The outermost loop consists of 2 loops* L_1 *(left) and* L_2 *(right). With* $O(S) = 1 + 1 \cdot (2 + O(L_1) + O(L_2))$ *and* $x(L_1) = 3$, $x(L_2) = 4$, *we obtain the individual loop overheads as* $O(L_1) = 1 + 3 \cdot (2 + O(3A) + O(4B))$ *and* $O(L_2) = 1 + 4 \cdot (2 + O(3C) + O(2D))$. *The innermost loops* (3A), (4B), (3C), (2D) *have the overheads* $1 + 6, 1 + 8, 1 + 6, 1 + 4$, *respectively. Hence,* $O(L_1) = 1 + 3 \cdot 18$ *and* $O(L_2) = 1 + 4 \cdot 14$, *and* $O(S)$ *becomes* 115 *cycles.*

In total, $T(S)$ of a given schedule S is defined as

$$\begin{aligned} T(S) &= (\sum_{i=1}^{|V|} (1 - flag(N_i)) \cdot L(N_i) \cdot q(N_i)) \\ &+ IO(S) + O(S) \end{aligned} \tag{10}$$

Hence, the explorer avaluates a three-dimensional cost function. For the following experiments, we characterize the influence of a chosen target processor by the following overhead parameters using the above target (overhead) functions:

- P_S: subroutine call overhead (number of cycles) (here: for simplicity assuming independence of actor, and no context to be saved and restored except PC and status registers).
- P_L: the number of program words for a complete loop instruction including initialization overhead.
- T_S: the number of cycles required to execute a subroutine call and a return instruction and to store and recover context information.
- T_L, T_L^{init}: loop overhead, loop initialization overhead, respectively in clock cycles.

Three real DSPs have been modeled, see Table 2. The DSP56k and TMS320C40 have high subroutine execution time overhead; the DSP56k, however, has a zero-loop overhead and high loop initialization overhead; and the TMS320C40 has a high loop iteration overhead but low loop initialization overhead.

System	Motorola DSP56k	ADSP 2106x	TI 320C40
P_L	2	1	1
P_S	2	2	2
T_L, T_L^{init}	0,6	0,1	8,1
T_S	8	2	8

Table 2: The parameters of 3 well-known DSP processors. All are capable of performing zero-overhead looping. For the TMS320C40, however, it is recommended to use a conventional counter and branch implementation of a loop in case of nested loops.

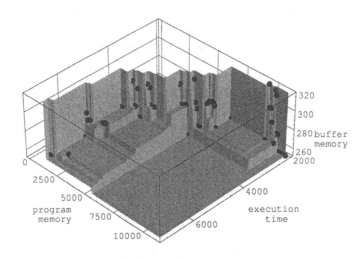

Figure 14: Motorola DSP56k

Example 4.8 *The CD to DAT example was taken as the basis to compare the design spaces of the above mentioned 3 different DSP processors. A design point is characterized by a string representing a periodic looped schedule. An Evolutionary Algorithm was developed as the supporting optimization algorithm, see also [21, 24, 26] for detailed results. The experimental results are visualized in Figures 14 for the Motorola DSP56k processor.*
The trade-offs between the 3 objectives are very well reflected by the extreme points. The rightmost points in the plots represent schedules that neither use looping nor subroutine calls. Therefore, there are optimal in the execution time dimension, but need a maximum of program memory because for each actor firing there is an inlined code block. In contrast, the leftmost points make excessive use of looping and subroutines which leads to minimal program memory requirements, however at the expense of a maximum execution time overhead. Another extreme point (not shown in the figures) satisfies $D(S) = 1021$, but has only little overhead in the remaining 2 dimensions. It stands for an implementation which includes the code for each actor only once by using inlining and looping. The schedule associated with this implementation is a single appearance schedule.

References

[1] T. Bäck, D. B. Fogel, and Z. Michaelevicz. *Handbook on Evolutionary Computation*. Inst. of Phy. Publ., Bristol, 1997.

[2] F. Balarin, A. Jurecska, and H. H. et al. *Hardware-Software Co-Design of Embedded Systems: The Polis Approach*. Kluwer Academic Press, Boston, 1997.

[3] A. Benveniste and G. Berry. The synchronous approach to reactive and real-time systems. *Proceedings of the IEEE*, 79(9):1270–1282, 1991.

[4] T. Blickle, J. Teich, and L. Thiele. System-level synthesis using Evolutionary Algorithms. *J. Design Automation for Embedded Systems*, 3(1):23–58, January 1998.

[5] J. Buck, S. Ha, E. Lee, and D. Messerschmitt. Ptolemy: A framework for simulating and prototyping heterogeneous systems. *International Journal on Computer Simulation*, 4:155–182, 1991.

[6] F. Cieslok, H. Esau, and J. Teich. EXPLORA- a tool for generic design space exploration. Technical report, TR No. 2/00, Computer Engineering Laboratory, University of Paderborn, Feb. 2000.

[7] L. Davis. *Handbook of Genetic Algorithms*, chapter 6, pages 72–90. Van Nostrand Reinhold, New York, 1991.

[8] D. Gajski, N. Dutt, A. Wu, and S. Lin. *High Level Synthesis: Introduction to Chip and System Design*. Kluwer, Norwell, Massachusetts, 1992.

[9] N. Halbwachs. *Synchronous Programming of Reactive Systems*. Kluwer Academic Publishers, Dordrecht, The Netherlands, 1993.

[10] D. Harel. Statecharts: A visual formalism for complex systems. *Science of Computer Programming*, 8, 1987.

[11] C. A. R. Hoare. *Communicating Sequential Processes*. Prentice Hall, Englewood Cliffs, NJ, 1985.

[12] S. Kirkpatrick, C. D. Gelatt, and M. P. Vecchi. Optimization by simulated annealing. *Science*, 220(4598):671–680, 1983.

[13] R. Lauwereins, M. Engels, J. A. Peperstraete, E. Steegmans, and J. V. Ginderdeuren. Grape: A CASE tool for digital signal parallel processing. *IEEE ASSP Magazine*, 7(2):32–43, Apr. 1990.

[14] E. Lee and D. Messerschmitt. Synchronous dataflow. *Proceedings of the IEEE*, 75(9):1235–1245, 1987.

[15] E. A. Lee and T. M. Parks. Dataflow Process Networks. Technical Report UCB/ERL 94/53, Dept. of EECS, UC Berkeley, Berkeley, CA 94720, U.S.A., 1994.

[16] E. A. Lee and T. M. Parks. Dataflow process networks. *Proceedings of the IEEE*, 83(5):773–799, 1995.

[17] S. Ritz, M. Pankert, and H. Meyr. High level software synthesis for signal processing systems. In *Proc. Int. Conf. on Application-Specific Array Processors*, pages 679–693, Berkeley, CA, 1992.

[18] R. Saracco, J. R. W. Smith, and R. Reed. *Telecommunications systems engineering using SDL*. North-Holland, Elsevier, Amsterdam, 1989.

[19] K. Strehl, L. Thiele, D. Ziegenbein, R. Ernst, and J. Teich. Scheduling hardware/software systems using symbolic techniques. In *Proceedings of the 7th International Workshop on Hardware/Software Codesign (CODES'99)*, pages 173–177, Rome, Italy, May 3–5, 1999.

[20] J. Teich. *Digitale Hardware/Software-Systeme: Synthese und Optimierung*. Springer-Lehrbuch, Heidelberg, New York, Tokio, 1997.

[21] J. Teich, E. Zitzler, and S. S. Bhattacharyya. 3D exploration of software schedules for DSP algorithms. In *Proc. CODES'99, the 7th Int. Workshop on Hardware/Software Co-Design*, Rome, Italy, May 1999.

[22] J. Teich, E. Zitzler, and S. S. Bhattacharyya. 3d exploration of uniprocessor schedules for DSP algorithms. Technical Report 56, Institute TIK, ETH Zurich, Switzerland, January 1999.

[23] P. P. Vaidyanathan. *Multirate Systems and Filter Banks*. Prentice Hall, 1993.

[24] E. Zitzler, J. Teich, , and S. S. Bhattacharyya. Evolutionary algorithm based exploration of software schedules for digital signal processors. In *Proc. GECCO'99, the Genetic and Evolutionary Computation Conference*, Orlando, U.S.A., July 1999.

[25] E. Zitzler, J. Teich, and S. Bhattacharyya. Multidimensional exploration of software implementations for DSP algorithms. *J. on VLSI Signal Processing*, 24:83–98, 2000.

[26] E. Zitzler, J. Teich, and S. Bhattacharyya. Evolutionary algorithms for the synthesis of embedded software. *IEEE Trans. on VLSI Systems*, to appear, 2000.

System Level Design
Using the SystemC Modeling Platform [1]

Joachim Gerlach
gerlach@informatik.uni-tuebingen.de

Wolfgang Rosenstiel
rosenstiel@informatik.uni-tuebingen.de

University of Tübingen • Department of Computer Engineering
Sand 13 • D-72 076 Tübingen • Germany

Abstract

This paper gives an overview of the SystemC modeling platform
and outlines the features supported by the SystemC class library.
The use of the modeling platform is shown in terms of an example.

1. Introduction

As system complexity increases and design time shrinks, it becomes extremely important that system specification be written down in a form that leads to unambiguous interpretation by the system implementers. The most common form of system specification, a written document, has several drawbacks: natural language is ambiguous and open to interpretation, the specification may be incomplete and inconsistent; and finally, there is no way to verify the correctness of such a specification. These drawbacks have driven many system, hardware, and software designers to create executable specifications for their systems. For the most part, these are functional models written in a language like C or C++. These languages are chosen for three reasons: first, they provide the control and data abstractions necessary to develop compact and efficient system descriptions, second, most systems contain both hardware and software and for the software, one of these languages is the natural choice, and third, designers are familiar with these languages and there exists a large number of development tools associated with them.

A functional model in C or C++ is essentially a program that when executed exhibits the same behavior as the system to be modeled. However, creating a functional model in a programming language like C or C++ is problematic because these languages are intended for software development and do not provide the constructs necessary to model timing, concurrency, and reactive behavior, which are all needed to create accurate models of systems containing both hardware and software. To model concurrency, timing, and reactivity, new constructs need to be added to C++. An object-oriented programming language like C++ provides the ability to extend a language through classes, without adding new syntactic constructs. A class-based approach is superior to a proprietary new language because it allows designers to use the language and tools they are familiar with.

1. This work was partially supported by the DFG Priority Research Program „Rapid Prototyping of Embedded Systems With Hard Time Constraints" and Synopsys Inc., Mountain View, CA.

R. Merker and W. Schwarz (eds.), System Design Automation, 27-34.
© 2001 *Kluwer Academic Publishers.*

2. The SystemC Approach

On September 1999, leading EDA, IP, semiconductor, systems and embedded software companies announced the "Open SystemC Initiative" (OSCI) and immediate availability of a C++ modeling platform called SystemC for free web download at the Embedded Systems Conference, San Jose, California. Achieving a break-through in industry cooperation, SystemC is the first result of the initiative, which enables, promotes, and accelerates system-level intellectual property (IP) model exchange and co-design using a common C++ modeling platform. Through an Open Community Licensing model, designers can create, validate, and share models with other companies using SystemC and a standard ANSI C++ compiler. In addition, electronic design automation (EDA) vendors have complete access to the SystemC modeling platform required to build interoperable tools. There are no licensing fees associated with the use of SystemC, and every company is free to join and participate. Backed by a growing community of well over 50 charter member companies, the Open SystemC Initiative includes representants from the systems, semiconductor, IP, embedded software and EDA industries. The steering group consists of a large number of market leading companies, including ARM, Cadence, CoWare, Ericsson, Fujitsu Microelectronics, Infineon Technologies, Lucent Technologies, Motorola, NEC, Sony Corporation, STMicro-electronics, Synopsys, Texas Instruments. The goal of the Open Community Licensing model is to provide a foundation to build a market upon, and the role of the steering group is to provide an environment of structured innovation ensuring that interoperability is retained.

Because of the excellent background of the Open SystemC Initiative, SystemC is on the best way to become a de-facto-standard for system-level modeling and design. This paper gives an overview of the SystemC modeling platform and outlines the features supported by the SystemC class library.

3. Overview of the SystemC Design Flow

The following overview refers to SystemC version 1.0 which is currently available for free web download at *www.systemc.org*.

The fundamental building blocks in a SystemC description are a processes. A process is similar to a C or C++ function that implements behavior. A complete system description consists of multiple concurrent processes. Processes communicate with one another through signals, and explicit clocks can be used to order events and synchronize processes. All building blocks are objects (classes) that are part of SystemC. Special data types required to model hardware efficiently are also provided as a part of the library. SystemC uses the full C++ language. A user only needs to understand how to use the classes and functions provided by the library, but she/he does not need to know how they are implemented. Using the SystemC library, a system can be specified at various levels of abstraction. At the highest level, only the functionality of the system may be modeled. For hardware implementation, models can be written either in a functional style or in a register-transfer level style. The software part of a system can be naturally described in C or C++. Interfaces between software and hardware and between hardware blocks can be easily described either at the transaction-accurate level or at the cycle-accurate level. Moreover, different parts of the system can be modeled at different levels of abstraction and these models can co-exist during system simulation. The use of C/C++ and the SystemC classes is not limited to the development of the system, but can also be used for the implementation of testbenches. The functionality of the SystemC classes together with the object-oriented nature of C++ provides a powerful mechanism for developing compact, efficient, and reusable testbenches. SystemC consists of a set of header files describing the classes and a link library that contains the simulation kernel. The header file can be used by the designer in her/his program. Any ANSI C++-compliant compiler can compile SystemC, together with the program. During linking, the SystemC library, which contains the simulation kernel is used. The resulting executable serves as a simulator for the system described, as shown in figure 1. Choosing C/C++ as the modeling language, a variety of software development tools like debuggers and integrated development environments can be utilized.

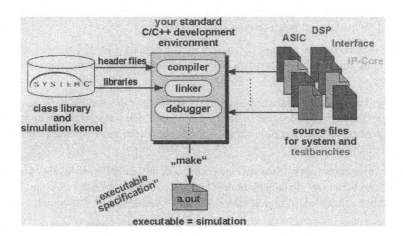

Figure 1. SystemC Design Flow.

4. Features of the SystemC Class Library

In the following, some features of SystemC version 1.0 are presented. More details can be found in the SystemC Release 1.0 Reference Manual which is included in the open source distribution of SystemC.

- Modules: In SystemC, the fundamental building block is a module. Processes are contained inside modules and modules support multiple processes inside them. Modules can also be used for describing hierarchy: a module can contain submodules, which allows to break complex systems into smaller more manageable pieces. Modules and processes can have a functional interface, which allows to hide implementation details and, for this, include blocks of IP.

- Processes: Processes are used to describe functionality. SystemC provides three different process abstractions to be used by hardware and software designers: methods (asynchronous blocks), threads (asynchronous processes) and clocked threads (synchronous processes).

- Ports: Ports of a module are the external interface passing information to and from a module, and triggering actions within the module. Ports can be single-direction or bidirectional.

- Signals: Signals create connections between module ports allowing modules to communicate. SystemC supports resolved and unresolved signals. Resolved signals can have more than one driver (a bus) while unresolved signals can only have a single driver.

- Signal types: To support different levels of abstraction, ranging from the functional level to the register-transfer level, as well as to support software, SystemC supports a rich set of signal types. This is different to languages like Verilog that only support bit and bit-vectors as signal types. SystemC supports both two-valued and four-valued signal types.

- Data types: SystemC has a rich set of data types to support multiple design domains and abstraction levels. The fixed precision types allow fast simulation. The arbitrary precision types can be used for computations with large numbers and to model large busses. SystemC supports both two-valued and four-valued data types. There is no size limitation for arbitrary precision SystemC data types. For example, SystemC version 1.0 provides arbitrary precision fixed-point data types, together with a rich set of overloaded operators, quantization and overflow modes, and type conversion mechanisms.

- Clocks: SystemC has the notion of clocks as special signals. Clocks are the timekeepers of the system during simulation. SystemC supports multiple clocks with arbitrary phase relationships.

- Reactivity: For modeling reactive behavior, SystemC provides mechanisms for waiting on events like clock edges and signal transitions. SystemC also supports watching for a certain event, regardless of the execution stage of the process (the most common example is the watching of a reset signal).

- Multiple abstraction levels: SystemC supports modeling at different levels of abstraction, ranging from high level functional models to detailed register-transfer level models. It supports iterative refinement of high level models into lower levels of abstraction.

- Cycle-based simulation: SystemC includes a cycle-based simulation kernel that allows high speed simulation. SystemC also provides mechanisms for simulation control at any point of the input specification.

- Debugging support and waveform tracing: SystemC classes have run-time error checking that can be turned on during compilation. The SystemC kernel contains basic routines to dump waveforms to a file (VCD, WIF, and ISDB format), which can be viewed by standard waveform viewers.

5. Example

In the following, the use of the SystemC modeling platform is shown in terms of an example. Objective of this quite small example is to give an overview of how systems are modeled using SystemC, so the reader should not concentrate on system functionality or system complexity. The example consists of two synchronous processes, *process_1* and *process_2*, communicating with one another. *process_1* increments the value of an integer input port by 5 and assigns the result to an integer output port, *process_2* increments the value of an integer input port by 3 and assigns the result to an integer output port. Both processes are connected in a way that an integer value is alternately incremented by *process_1* and *process_2*. Process synchronization is done via boolean signals (see figure 2).

Figure 3 shows the header and the implementation code file of *process_1*. The process is encapsulated in a SystemC module given by a C++ class. This can be done using the SystemC macro *SC_MODULE* (*SC_MODULE(process_1)* {...} is equivalent to *struct process_1: sc_module {...}*, where the base class *sc_module* is provided by the SystemC class library). Module ports are described by data members of the C++ class. Input ports are declared by

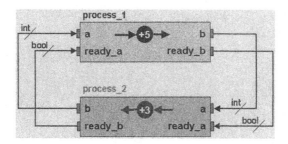

Figure 2. Example system.

sc_in<T>, output ports are declared by *sc_out<T>* (where *T* is an arbitrary data type). In addition, a special input port of type *sc_in_clk* has to be specified for the clock signal. The process functionality is encapsulated in a function member *void do_process_1()* of the C++ class. *process_1* is implemented to be a synchronous process, which is done in the class constructur. Again, the class constructur doesn't have to be coded directlty, but can be specified using the SystemC macro *SC_CTOR*. In the body of the class constructor, *process_1* is made to be a synchronous process by using *SC_CTHREAD*. Arguments of SC_CTHREAD are the name of the process function member and the clock edge to which the process is sensitive. In the body of the function member, *process_1* waits until the boolean input signal *ready_a* becomes true. Then, input port *a* is read and the corresponding integer value is assigned to a local integer variable *v*. *v* is incremented by 5, displayed for evaluation purpose, and assigned to output port *b*. Next, the boolean input signal *ready_b* is set to true for one clock cycle.

```
// header file: process_1.h

SC_MODULE( process_1 ) {

    // Ports
    sc_in_clk clk;
    sc_in<int> a;                          // implementation file: process_1.cc
    sc_in<bool> ready_a;
    sc_out<int> b;                         #include "systemc.h"
    sc_out<bool> ready_b;                  #include "process_1.h"

    // Process functionality                void process_1::do_process_1()
    void do_process_1();                    {
                                                int v;
    // Constructor
    SC_CTOR( process_1 ) {                      while ( true )
        SC_CTHREAD( do_process_1 , clk.ps() );  {
    }                                               wait_until( ready_a.delayed() == true );
                                                    v = a.read();
};                                                  v += 5;
                                                    cout << "P1: v = " << v << endl;
                                                    b.write( v );

                                                    ready_b.write( true );
                                                    wait();
                                                    ready_b.write( false );
                                                }
                                            }
```

Figure 3. *process_1.h* **and** *process_1.cc.*

process_2 is specified in a very similar way as shown in figure 4.

```
// header file: process_2.h

SC_MODULE( process_2 ) {

    // Ports
    sc_in_clk clk;
    sc_in<int> a;
    sc_in<bool> ready_a;
    sc_out<int> b;
    sc_out<bool> ready_b;

    // Process functionality
    void do_process_2();

    // Constructor
    SC_CTOR( process_2 ) {
        SC_CTHREAD( do_process_2 , clk.ps() );
    }

};
```

```
// Implementation file: process_2.cc

#include "systemc.h"
#include "process_2.h"

void process_2::do_process_2()
{
    int v;

    while ( true )
    {
        wait_until( ready_a.delayed() == true );
        v = a.read();
        v += 3;
        cout << "P2: v = " << v << endl;
        b.write( v );

        ready_b.write( true );
        wait();
        ready_b.write( false );
    }
}
```

Figure 4. *process_2.h* **and** *process_2.cc.*

The next step is to create an instance of each process and tie them together with signals in a top-level routine. By convention, this routine is called *sc_main*. Figure 5 shows the corresponding source code file. In the top-level routine, all process header files and the file *systemc.h* are included, because these files contain the declaration for all the process classes and SystemC library functions. Inside *sc_main*, signals used for process communication are declared. After the signals are instantiated, the clock object, which is a special signal, is instantiated. In our example, the name of the clock object is Clock, it has a period of 20 time units and a 50% duty

```
// implementation file: main.cc

#include "systemc.h"
#include "process_1.h"
#include "process_2.h"

int sc_main (int ac,char *av[]) {

    sc_signal<int> s1 ( "Signal-1" );
    sc_signal<int> s2 ( "Signal-2" );
    sc_signal<bool> ready_s1 ( "Ready-1" );
    sc_signal<bool> ready_s2 ( "Ready-2" );

    sc_clock clock( "Clock" , 20 , 0.5 , 0.0 );

    process_1 p1 ( "P1" );
    p1.clk( clock );
    p1.a( s1 );
    p1.ready_a( ready_s1 );
    p1.b( s2 );
    p1.ready_b( ready_s2 );
```

```
    process_2 p2 ( "P2" );
    p2.clk( clock );
    p2.a( s2 );
    p2.ready_a( ready_s2 );
    p2.b( s1 );
    p2.ready_b( ready_s1 );

    if ( (argc == 2) &&
         (strcmp (argv[1],"w") == 0) ) {

        sc_trace_file *tf
            = sc_create_wif_trace_file( "trace_file" );
        sc_trace( tf,s1,"Signal-1" );
        sc_trace( tf,s2,"Signal-2" );
        sc_trace( tf,ready_s1,"Ready-1" );
        sc_trace( tf,ready_s2,"Ready-2" );
    }

    s1.write( 0 );
    s2.write( 0 );
    ready_s1.write( true );
    ready_s2.write( false );

    sc_start(100000);

    return 0;
}
```

Figure 5. Top-level routine *main.cc.*

cycle. Next, processes *process_1* and *process_2* are declared and the ports are connected by signals. *sc_main* also contains the declaration of an output file for waveform tracing and the specification of a set of signals to be traced during simulation. In our example, waveform tracing is optional and can be activated with the *w* argument during the call of the executable. Furthermore, an initialization of the signals is done. Once all processes are instantiated and connected to signals, the clock is generated to simulate the system. This is done by the SystemC function *sc_start(n)* where *n* is the number of time units for which the simulation is intended to last.

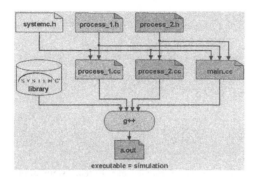

Figure 6. Source code file structure.

In our example, the entire system consists of three implementation files (*process_1.cc*, *process_2.cc*, *main.cc*) and two header files (*process_1.h*, *process_2.h*). Figure 6 shows the file structure of the source code. The implementation files can be compiled individually and finally linked with the SystemC library. In addition, compilation of each file requires header files from SystemC. Compilation can be done using any standard ANSI C++ compiler (for example, gnu *gcc*). Executing the resulting binary is equivalent to running a simulation of the system description for the specified number of time units.

In our example, system behavior can be verified by observing the process' outputs. Figure 7 shows the first view output lines produced when executing the binary.

```
SystemC (TM) Version 1.0 — Apr 4 2000 10:12:32
          ALL RIGHTS RESERVED
    Copyright (c) 1988-2000 by Synopsys, Inc.

P1: v = 5
P2: v = 8
P1: v = 13
P2: v = 16
P1: v = 21
P2: v = 24
P1: v = 29
P2: v = 32
P1: v = 37
P2: v = 40
P1: v = 45
P2: v = 48
P1: v = 53
P2: v = 56
P1: v = 61
P2: v = 64
P1: v = 69
P2: v = 72
```

Figure 7. Process outputs.

Figure 8 shows parts of the waveform file produced during simulation. The execution of the binary (which means, the simulation of the system description for 100000 time units) takes 0.08 seconds on a Sun Ultra Sparc 5 with 384 MByte main memory (for runtime measurement, the process outputs and writing of waveform files were skipped). Using version 0.9 of the SystemC library, the simulation of the system description for 100000 time units takes 0.31 seconds on the same machine.

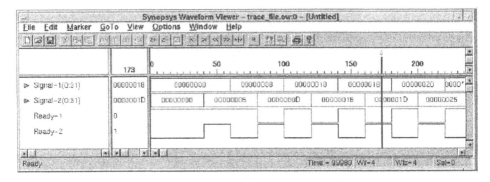

Figure 8. Waveform view of a SystemC simulation.

6. Conclusion

This paper gives a brief overview of the SystemC modeling platform. SystemC provides innovative mechanisms for C++-based system-level description and is freely available through an open source licensing model. Because of those facts and the continuously growing number of leading EDA, IP, semiconductor, systems and embedded software companies joining the Open SystemC Initiative, SystemC is on the best way of becoming a de-facto-standard for system-level specification.

Requirements for Static Task Scheduling in Real Time Embedded Systems

Chun Wong*, Filip Thoen, Francky Catthoor†, Diederik Verkest
IMEC, Kapeldreef 75, B-3001, Leuven, Belgium
{chwong, catthoor, verkest}@imec.be
*Also Ph.D. student of Katholieke Univ. Leuven-ESAT
†Also professor of K.U.Leuven-ESAT

Abstract

Static task scheduling is an important step in embedded system design. In this paper, we have studied a recently proposed representation model, called *Multi Thread Graph* and a given static scheduling heuristic based on this *MTG*. Applying the heuristic to the software part of a *GPS* receiver, we have successfully scheduled the tasks under the required timing constraints. The experiment shows that extension to the *slack* concept is effective as a scheduling metric in systems where timing constraints are the only issue. However, by analyzing the results we have also identified other cost related factors to be incorporated in the original heuristic to arrive at an applicable task scheduling algorithm.

Keywords: Multi Thread Graph, static task scheduling, slack, real time, embedded system.

1. Introduction

The design of embedded systems spans at least two abstraction levels, namely, *operation (instruction) level* and *task level*. The operation-level design takes care of the coding of detailed behaviors, e.g., the interface synthesis or the code generation for the on-chip programmable components. On a higher level, which we call the task-level, an embedded system can be viewed as a number of interacting concurrent tasks. The major challenge on this level is to coordinate the interactions between tasks and the interactions between the system and its environment in a cost-effective and real-time way.

Task scheduling is a major step of the task-level design. It takes care of the processor assignment and orders the execution of tasks on each processor. Scheduling is done under real-time constraints with cost considerations, like the processor energy cost, the energy cost for data storage and transfer, etc.

In this paper, we restrict our discussion to a single processor context. Different execution orders directly affects meeting timing constraints. Consequently, how to schedule these tasks on one processor is still a non-trivial job in the global system design trajectory [1].

Two ways exist to tackle the scheduling problem. The conventional way is to use *RTOSs* (Real Time Operating Systems). Unfortunately, an RTOS only provides limited functionality and hence designers have to tune the application code or task priorities to meet timing constraints. Secondly, an RTOS assumes a specific processor and a particular I/O configuration. Consequently, porting to other platforms usually requires partial code revision. Finally, an RTOS is a stripped operating system, which trades generality for cost efficiency. But even the runtime overhead of an RTOS may become unacceptable when hard real-time performance is of prime importance.

Contrary to the above approach, it is more promising to synthesize an application specific runtime kernel tuned to the concurrent task behavior. Such a design flow involves taking a high-level system specification together with its timing constraints and generating an

R. Merker and W. Schwarz (eds.), System Design Automation, 35-44.
© 2001 *Kluwer Academic Publishers.*

application specific executable program, which obeys these constraints. Since most of the concurrency has already been removed by preceding steps at design-time in this approach, the timing behavior is improved and the run-time overhead, including code size/energy overhead, is reduced. It is a model-based methodology. In a limited context, i.e., the software-related context, this approach reduces to *software synthesis* [2, 3]. *Static task scheduling* is a major step in this scenario.

The focus of this paper is to study some important issues and requirements in *static task scheduling*. It is organized as follows. Section 2 summarizes the current research on modeling and task scheduling. Section 3 introduces the *MTG* model. Based on this model, the static *slack*-driven scheduling heuristic is discussed in Section 4. Applying the heuristic to the embedded software of the *GPS* receiver is discussed in Section 5. Section 6 analyzes and discusses the problems found in the experiment. Section 7 concludes the paper with interesting future extensions of the static scheduling principle.

2. Related Work and Motivation

2.1 System Specification Model

Specification models are the means to explore task scheduling. A comprehensive survey of formal specification models of real-time systems is given in [4]. From the overview, it can be concluded that
   1. They focus on a too low abstraction level to perform *software synthesis*;
   2. They typically only allow periodic processes and no complex process control constructs;
   3. They lack or only have limited support for timing constraints.
In contrast, the *MTG* model [5] tackles the above shortcomings by coupling traditional *CDFG* (Control Data Flow Graph) related information to a specific *Petri Net* instantiation. While the *CDFG* is well suited for capturing fine-grain behavior, data communication and local control, *Petri Net* offers powerful support for the high level task concurrency information, i.e., global control and timing specification. In addition, *MTG* first makes a proper selection of non-deterministic behaviors to be included. Then it models them explicitly with *event nodes* and *semaphore nodes*. Secondly, it introduces *constraint edges* to model timing constraints consistently.

2.2 Task Scheduling Algorithms

In the real-time community, researchers use a *black-box* view of the task behavior. Comprehensive overviews of scheduling algorithms for real-time systems are given in [5, 6, 7, 8]. In contrast, in the embedded system community, many papers focus on *white-box* task descriptions, which are typically not available at the early design stage.

Scheduling algorithms are roughly divided into dynamic and static scheduling. In a multiple processor context, when the application has a large amount of non-deterministic behavior, dynamic scheduling has the flexibility to balance the computation load of processors at run-time. However, the run-time overhead for code size and power consumption may be excessive. In addition, most DSP applications have limited non-deterministic behavior. Moreover, it is usually impossible to make globally optimal scheduling decisions at run-time. Consequently, scheduling decisions should be made as many as possible at design-time.

A large body of scheduling algorithms stems from the *schedulability test* proposed in [9]. In this paper, *critical instant, fixed priority, dynamic priority, Rate Monotonic Scheduling* and *Earliest Deadline First Scheduling* were also introduced. All these concepts are the foundations for the successive work until today. The resulting scheduling algorithms are restricted to the single processor and *periodic* task context. Based on the *schedulability test*, P. Altenbernd [11] proposes *Deadline Monotonic Scheduling* for parallel communicating *tasks*. It reduces the schedulability test to the automatic computation of deadlines and offsets.

Since more and more embedded systems are targeted at multiple processor architectures, multiple processor scheduling plays a more and more important role. H. El-Rewini *et al* [12] give a clear introduction to the task scheduling in multiprocessing systems. C. J. Hou *et al* [13] alleviate the *saturation effect* [14] caused by excessive inter-processor communication in distributed embedded systems. However, their work does not address the problem of allocating *tasks* to *processing nodes*, i.e., the tremendous search space to explore. J. M. Rabaey *et al* [15] try to maximize the throughput by balancing the computation load of the distributed processors. All the partition and scheduling decisions are made at compile-time, so the approach is a *static* one. The approach is limited to pure data flow applications. W. Wolf *et al* [10, 16] address meeting hard real-time constraints by avoiding preemption. The key point of the algorithm is to compare the execution times of two *tasks* to determine whether to preempt or execute one of the *tasks*.

Only a few scheduling papers consider power issue. Currently all this work takes a *white-box* view. For example, M. Potkonjak *et al* [17] propose a variable voltage technique. It dynamically changes voltage according to whether the deadline is relaxed or tight. Since the technique demands the detailed task behavior for the schedulability test, it is meant for lower abstraction levels.

2.3 Conclusions

Though many basic scheduling kernels exist, but none of them does a good job on its own for a complex application. They are based either on a too low abstraction level, which is termed as *white-box* model in [5], or a *black-box* model. While the former lacks task-level information, the latter fails to capture cost-effective information. The fact drives us to develop a scheduling algorithm based on a *grey-box* model [5]. The algorithm will be a new combination of basic scheduling methods. Since the problem of finding an optimally valid task ordering is NP-complete in the strong sense [18], we resort to heuristic strategies. Though heuristic approaches are blamed for potential sub-optimal solutions, we can discover the really important factors in task scheduling by applying the heuristic to real life examples.

3. *MTG* Model Introduction

Our *slack*-based scheduling algorithm is built on the *MTG* model proposed in [5]. However, the focus of this paper is not the modeling issue. In addition, *MTG* is a very comprehensive model. Hence, in this section we only introduce the *MTG* model for the sake of understanding the rest of the paper.

Generally speaking, *MTG* is a combination of a specific *Petri Net* instantiation and a more traditional *CDFG* model. As a result, the operational semantics of an *MTG* resembles that of a *Petri Net*. An *MTG* mainly consists of two groups of elements, namely, *nodes* and *edges*. **Fig. 1** graphically illustrates the *MTG* model.

Nodes can be divided into two basic types, which are *operation node* and *data node* respectively. *Operation nodes* can be further divided into *behavior node* and *control flow node*. *Behavior nodes* represent data operations, including *thread node* and *hierarchical thread node*. *Control flow nodes* include *or*, *and*, *source*, *sink*, *event*, *synchronization* and *semaphore* nodes. *Data nodes* represent either data transactions or communications. *Data nodes* include *system I/O node* and *local shared memory variable node*.

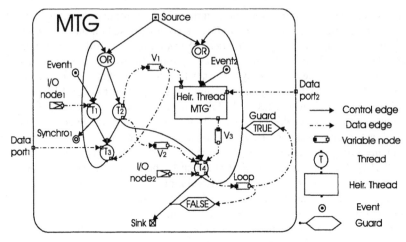

Fig. 1 Example of *Multi Thread Graph*

Three types of *edges* exist, which are *control*, *data* and *constraint edges*. When extended with time notion, *MTG* uses *constraint edges* to express the timing constraints. Timing constraints can be latency constraints, response time constraints, and rate constraints.

The key concept of *MTG* is the *thread node*. By definition, A thread node T_i is a maximum set of connected operations with deterministic execution latency $\Lambda(T_i) = [\delta_i, \Delta_i]$. δ_i and Δ_i are the minimum and maximum execution time of the thread node T_i. A number of thread nodes can be grouped together as a *hierarchical thread node*. However, the hierarchical thread node is just defined for reuse purpose. Instead, the heavily communicating and synchronizing thread nodes are clustered into a *task*. A *task* has the characteristic that it will execute autonomously once it is triggered. In other words, it will not need synchronization from the outside.

After *task clustering*, we have a number of communicating *tasks*, which are triggered by events *dynamically*. Inside a *task*, we *statically* schedule the *thread nodes* under intra-task timing constraints derived from the system specification. Scheduling the *tasks* at run time is left to the run time executive. Global timing constraints spanning different *tasks* are not taken into account yet. The purpose of introducing such a two-layer division, namely, *thread node* and *task*, is to allow ordering software segments as much as possible at design time. Consequently, this reduces the run time overhead to the lowest degree.

In summary, the *MTG* model captures both fine-grain (i.e., operation-level) behavior and high-level information. To make a good system specification, both types of information are required.

4. *Slack*-based Static Task Scheduling Heuristic

4.1 Calculation of the Maximum Time Separation and *Slack*

Most timing analysis problems boil down to the calculation of maximum separation time between two nodes in a graph. This also applies to *slack* calculation. To calculate the maximum time separation between two nodes, we have adapted an existing algorithm, TSE (Time Seperation of Events) [19]. The adapted algorithm calculates two values for each node o_j with respect to o_{from}, denoted by $m(o_j)$ and $M(o_j)$. $m(o_j)$ is defined as follows.

$$m(o_j) = \max\{\delta(p)| \forall p : o_j \xrightarrow{\ p\ } o_{from}\} \tag{1}$$

with

$$\delta(p) = \sum\{\delta(o_k)| o_k \text{ is a node on path } p\} \tag{2}$$

The value $m(o_j)$ is defined as the maximum accumulated sum of lower execution latencies among all the paths from o_j to o_{from}. If there is no path from o_j to o_{from}, we assign $m(o_j) = 0$. For understanding the rest of the paper, it is enough to know that $M(o_j)$ is computed in a forward topological order. The maximum separation between o_{from} and o_j is then

$$sep^{max}(o_{from}, o_j) = M(o_j) - m(o_j) \tag{5}$$

The *slack* of a node o$_i$, denoted by $sl(o_i)$, is defined as the maximum delay which can be inserted before the start of o$_i$ such that no timing constraint is violated. The less *slack* a node has, the more urgent the node is and hence the earlier the node should be scheduled. The slack of node o_j in the presence of a maximum constraint $\varepsilon_{l,i}$ (between nodes o_l and o_i) with value $w_{l,i}$ is given by the following backward recursive formula:

$$\begin{cases} sl(o_i) = |w_{l,i}| - sep^{max}(o_i, o_i) - \Delta_l \\ sl(o_j) = \min_{o_k \in successor(o_j)}\{sl(o_k) + sep^{max}(o_i, o_k) - sep^{max}(o_i, o_j) - \Delta_j\} \\ sl(o_x) = +\infty, \text{ if } \forall p : o_x \xrightarrow{\ p\ } o_l \Rightarrow o_i \in p \end{cases} \tag{6}$$

The formula decomposes the *slack* calculation into two major steps, namely the maximum separation calculation and the back traversal of *slack* calculation.

4.2 Scheduling Algorithm

Algorithm 1 greedily schedules a given *task* with a node chosen as its root node, denoted by *ploar(task, o$_{root}$)*. The *task* is serialized with respect to the already constructed partial odering at each interation. This ordering is constructed incrementally until no unscheduled nodes are left. Consequently, a full order results.

Algorithm 1
begin
    $ord(task) \leftarrow \langle o_{root} \rangle$
    $o_{current} \leftarrow o_{root}$
    while (unordered candidate nodes) **do**
        $task_s \leftarrow serialize\,(task, ord(task))$ /* serialize task with partial ordering $ord()$ */
        $Scand \leftarrow compatible\,(task_s)$
        $o_{new} \leftarrow \min_{o_i \in Scand}\{f(o_i)\}$ /* select minumum slack candidate */
        $ord(task) \leftarrow \langle ord(task), o_{new} \rangle$
        if (negative slacks) **then return** *invalid* ordering
        $o_{current} \leftarrow o_{new}$
    end while
    return $ord(task)$ /* return valid ordering */
end

The routine $compatible\,(task_s)$ returns the set of candidate nodes with respect to the partial ordering $task_s$ to be scheduled next. To define it, the original task graph is first augmented with the (constructed) partial ordering $ord(task) = \langle o_1, o_2, \cdots, o_k \rangle$ by the *serialize()* routine. This routine adds the relations $\{(o_i, o_j) \mid i \leq j \leq k\}$ to the original task control structure. The priority function $f(o_i)$ is based on the *slack* metric discussed above. By selecting the node with the minimum *slack*, the most urgent nodes are scheduled first and most chance exists that a valid ordering results.

5. Real-life Application of the Slack-Driven Scheduling Heuristic

Real-time embedded systems span a wide range. Among them, we have chosen the *GPS* receiver as our initial test vehicle. First, it is simple in that it consists of a reasonable amount of C code. Secondly, it contains a reasonable amount of concurrency.

5.1. GPS Receiver and its MTG Representation

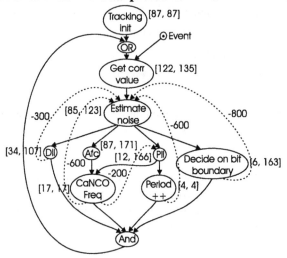

Fig. 2 *MTG* of the tracking loop with control and constraint edges

The *GPS* receiver in this study mainly consists of a hardware module and a software module. The software module processes the data provided by the hardware module according to some algorithm. In the software module, the interesting part for testing a task scheduling heuristic is the tracking loop because it contains partially ordered *thread nodes* and *constraint edges* between them. Because data communication is not taken into account during scheduling at present, to avoid cluttering the picture, we show the *MTG* of the tracking loop only with control and constraint edges in **Fig. 2**.

In Fig. 2, interrupts from the hardware module are modeled by *event* nodes. They inform the DSP processor that a new set of data is available. Annotated on the graph is the execution latency of every node. To keep consistent with execution latencies, we express timing constraints in clock cycles.

5.2 Scheduling Results

We summarize the scheduling procedure with the following tables and graphs. In step 1, we calculate the *slack* values of every node after node *Estimate noise*, which are listed in **Table 1**. Since node *Dll* has the minimum *slack*, it is scheduled immediately after node *Estimate noise*.

As a result, we get **Fig. 3**. Repeating Step 1 for four times, we get the final execution order of all the nodes as shown in **Fig. 4**.

	Dll	CaNCOFreq	Period++	Afc	Pll	Decide on bit boundary
Slack	70	289	307	289	294	514

Table 1 Slack values calculated in Step 1

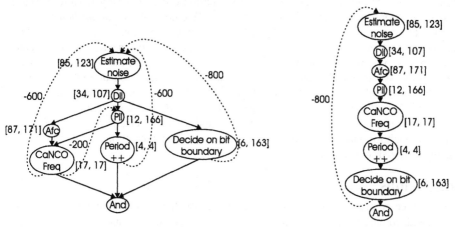

Fig. 3 Partial order after Step 1   **Fig. 4** Partial order after Step 5

	Decide on bit boundary
Slack	49

Table 2 Slack values calculated in Step 6

Step 6 calculates the *slack* of node *Decide on bit boundary*. Since it is a positive value, no timing constraint is violated at the end of scheduling. In other words, we have obtained a fully sequential execution order, which meets the specified timing constraints.

6. Analyses and Discussions

With the proposed heuristic, we have successfully scheduled the concurrent *thread nodes* in the software module of a *GPS* receiver. However, through this experiment, we have identified the following issues to be further investigated.

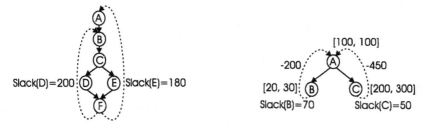

Fig. 5 Example of different start nodes   **Fig. 6** Effects of execution time

First, the *slack* of a node represents urgency with respect to the start node of a specific timing constraint. Consequently, different nodes can be compared with *slacks* only when they share a common start node. Considering the example in **Fig. 5**, which one of node D and E is more urgent? Because the urgencies refer to different start nodes, it makes little sense to perform such a comparison.

Secondly, *slack* describes only the urgency of a node in meeting a timing constraint. Sometimes other metrics like the execution time needs to be taken into account during scheduling. In **Fig. 6**, though the *slack* of C is less than that of B, scheduling C before B will obviously violate the timing constraint between A and B. In contrast, if B is scheduled before C, timing constraints will be met.

Thirdly, let us look at Fig. 3 again, which is the partial order after step 1. Here we have three candidate nodes to be scheduled after *Dll*. If we leave node *Decide on bit boundary* out temporarily for the moment, between *Afc* and *Pll*, which one should we choose? *Slack* calculation shows that *Afc* has a less *slack*, so it is chosen as the next node to execute. However, in this example, execution latencies are measured from the real set-up of the *GPS* receiver. Let us assume that *Pll* has a *slack* less than *Afc*, an invalid scheduling will happen. Only at a later stage when a negative *slack* of *CaNCOFreq* is detected, we can backtrack and schedule *Afc* before *Pll*, though *Afc* has a larger *slack* than *Pll*. This complexity is due to the fact that when a node is scheduled, *slacks* of other nodes will be changed. So the *slack* by itself is not enough for making a good scheduling decision. The interaction of different timing constraints also needs consideration.

7. Conclusions

From the three points discussed above, we draw the conclusion that an integral scheduling algorithm needs to incorporate, in addition to *slack*, execution latencies and interaction among entangled timing constraints.

Up to now, there is also no cost model for any of the *black-box* task scheduling algorithms, which are the only ones aimed at the high abstraction level where we also focus on. However, three kinds of costs exist, which could and should be taken into account. In the design of embedded systems, the data transfer and storage related cost, i.e., the background memory cost is largely defined already during task-level *Data Transfer and Storage Exploration* [20], which is a stage that should precede *static task scheduling* in a global top-down design flow [1]. Intra-task scheduling also influences the foreground memory, i.e., registers and register files, which forms the second cost factor. The third cost factor comes from units in the data path, like the number of ALUs. It may be difficult to estimate the last two costs at the task level because they are only fixed at the end of the instruction-level concurrency management stage. Still, they should be at least estimated in a crude form for relative comparisons. An integral cost function is by no means trivial, in that it not only evaluates the optimality of a scheduling result but also provides guidance during scheduling. This is the topic of current research.

Finally, as mentioned in Section 3, the heuristic doesn't yet take into account global timing constraints at present. This problem also needs to be tackled to construct a practically useful task scheduling algorithm on the *grey-box* level.

Acknowledgement

It is our pleasure to thank Johan Cockx for stimulating discussions, and Hans Cappelle and Paul Coene for providing the C code of the *GPS* receiver.

References

[1] F. Catthoor, D. Verkest, E. Brockmeyer: Proposal for unified system design meta flow in task-level and instruction-level design technology research for multi-media applications, Proceedings of 11th International Symposium on System Synthesis, Taiwan, 1998, 89-95

[2] M. Chiodo et al: Synthesis of software programs for embedded control applications, 32nd Design Automation Conference Proceedings 1995, ACM, New York, 1995

[3] P. Chou et al: Software architecture synthesis for retargetable real-time embedded systems, In Proceedings of 5th International Workshop on Hardware/Software Codesign, Braunschweig, Germany, March 1997

[4] J. S. Ostroff: Formal methods for the specification and design of real-time safety critical systems, Journal of Systems and Software (April 1992) 33-60

[5] F. Thoen; F. Catthoor: Modeling, Verification and Exploration of Task-Level Concurrency in Real-Time Embedded Systems, Kluwer Academic Publishers, 1999

[6] L. Sha; R. Rajkumar; J. Lehoczky: Priority inheritance protocols: An approach to real-time synchronization, IEEE Trans. On Computers, Vol. 39(9) (September 1990) 1175-1185

[7] K. Ramamotitham; J. A. Stankovic: Scheduling algorithms and Operating Systems Support for Real-Time Systems, Proc. IEEE, Vol. 82(1) (January 1994) 55-67

[8] N. Audsley; A. Burns; R. Davis; K. Tindell; A. Wellings: Fixed priority preemptive scheduling: An historical perspective, Journal of Real-time Systems, Vol. 8(2) (1995) 173-198

[9] C. L. Liu et al: Scheduling Algorithms for Multiprogramming in a Hard-Real-Time Environment, Journal of the Association for Computing Machinery, Vol. 20 (January 1973) 46-61

[10]   Y. Li; W. Wolf: Hierarchical Scheduling and Allocation of Multirate Systems on Heterogeneous Multiprocessors, Proceedings of European Design and Test Conference, ED & TC 97, IEEE Computer Soc. Press, Los Alamitos, CA, USA, 1997, 134-139

[11]   P. Altenbernd: Deadline-monotonic Software Scheduling for the Co-synthesis of Parallel Hard Real-Time Systems, Proceedings of the European Design and Test Conference ED&TC, IEEE Computer Soc. Press, Los Alamitos, CA, USA, 1995, 190-195

[12]   H. El-Rewini et al: Task Scheduling in Multiprocessing Systems, IEEE Computer (December 1995) 27-37

[13]   C. J. Hou et al: Allocation of Periodic Task Modules with Precedence and Deadline Constraints in Distributed Teal-Time Systems, IEEE Transactions on Computers, Vol. 46, No. 12 (December 1997) 1338-1356

[14]   P. Y. Ma et al: A Task Allocation Model for Distributed Computing Systems, IEEE Transactions on Computers, Vol. 31, No. 1, (January 1982) 41-47

[15]   P. D. Hoang; J. M. Rabaey et al: Scheduling of DSP Programs onto Multiprocessors for Maximum Throughput, IEEE Transactions on Signal Processing, Vol. 41, No. 6 (June 1993) 2225-2235

[16]   Y. Li; W. Wolf: Scheduling and Allocation of Single-Chip Multiprocessors for Multimedia, 1997 IEEE Workshop on Signal Processing Systems SiPS 97 Design and Implementation Formerly VLSI Signal Processing IEEE, New York, NY, USA, 1997, 97-106

[17]   I. Hong; M. Potkonjak et al: Power Optimization of Variable Voltage Core-Based Systems, Proceedings of 1998 Design and Automation Conference 35th DAC, IEEE, New York, NY, USA, 1998, 176-181

[18]   M. Garey; D. Johnson: Computers and Intractability, San Francisco, CA, W.H. Freeman Company, 1979

[19]   H. Hulgaard: Timing Analysis and Verification of Timed Asynchronous Circuits, Ph.D. thesis, Univ. of Washington, 1995

[20]   F. Catthoor *et al*: Custom Memory Management Methodology, Kluwer Academic Publishers, 1998

Relocalization of Data Dependences in Partitioned Affine Indexed Algorithms

Uwe Eckhardt   René Schüffny   Renate Merker

eckhardt(schueffn)@iee.et.tu-dresden.de   merker@iee1.et.tu-dresden.de

Institute of Circuits and Systems, TU Dresden

Abstract

Localization is an essential step in the array synthesis which leads to communications occuring only between neighboring processor modules. Costs of localization are an increase of allocated memory and total amount of communications within the architecture. In general, localization is applied to the whole algorithm before or in mutual dependence with scheduling and allocation. If resource constraints are taken into consideration, then scheduling and allocation are modified by several partitioning methods. If partitioning has been applied, localization should be revised due to the fact that partitioning makes localization partially unnecessary. Synthesis from completely localized code leads to very inefficient utilization of the memory system, to high implementation and communication costs, and hence, to high power consumption. Relocalization gives an approach to restrict localization to parts of partitioned algorithms where it is necessary.

1   Introduction

In many applications with severe timing and power constraints and high data throughput the main computation expenditure stems from sub-algorithms which belong to the class of affine indexed algorithms (AIA). These sub-algorithms are often encoded with nested loops or with fan-in operators like $\sum, \prod, max$, and gcd and represent a large sets of operations. For AIAs efficient solutions exist for the allocation (module selection combined with binding) and scheduling tasks with a complexity which is independent of the number of operations. These synthesis algorithms are capable to synthesize architectures with piecewise regular processor arrays which can cope with the severe timing constraints.

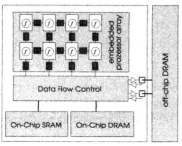

Figure 1: *Example of a target architecture*

First treatises on automatons with space-time-structure date from the 60's and 70's, e.g. by McCluskey [8] and von Neumann [13]. Since the advent of VLSI, many array architectures for signal processing algorithms and approaches to a systematic array synthesis for several algorithm classes have been proposed, e.g.[6, 7, 14, 9, 10, 17]. Todays DSPs and VSPs contain array architectures.

We consider the design of systems with embedded processor arrays and memory hierarchy at system level, i.e. the behavioral description of the design is given by algorithms and the structural description is given by processor modules, memory modules, busses, and controllers. **Fig.**1 shows an example of a target ar-

R. Merker and W. Schwarz (eds.), System Design Automation, 45-55.

chitecture. The first step in a synthesis of a processor array for a given affine indexed algorithm is an embedding and localization of data dependences, e.g.[17, 16]. The resulting algorithm has piecewise regular data dependences. Thereafter, allocation and scheduling are defined by a space-time-partitioning of the piecewise regular algorithm. Subsequently, a co-partitioning of the algorithm is performed to adjust the array topology, the capacity of local memories and the I/O-behavior between array and peripheral memory system [5]. Object of a localization of data dependences is that communications occur only between neighboring processor modules in the resulting array architecture. Costs of this localization are an increase of the number of variables in the algorithm code and hence, an increase of necessary capacities of memory modules and an increase of data transfers on busses between architecture modules.

Co-partitioning is a combined application of LSGP- and LPGS-partitioning. If LSGP-partitioning has been applied, then localization within partitions can be omitted due to the locally sequential processing within partitions. In the case of an LPGS-partitioning, a localization of data dependences between partitions can be omitted due to a sequential processing of partitions.

Object of our approach is a localization which takes the applied partitioning into account to avoid unnecessary large memory capacities and unnecessary data transfers.

The paper is organized as follows: Affine indexed algorithms are defined in Section 2. In Section 3, we show the embedding and localization of data dependences and the resulting coding as piecewise regular algorithm. In Section 4, we consider scheduling and allocation for the piecewise regular algorithm without resource constraints. Section 5 describes different algorithm partitioning schemes. In Section 6, we present an approach to localize data dependences of partitioned algorithms.

2　Affine Indexed Algorithms

An affine indexed algorithm is a set of statements

$$y_j[l_j(\mathbf{o})] := \underset{\mathbf{h}\in\mathcal{H}_j}{\Phi} f(y_i[r_{i,j}(\mathbf{o}+\mathbf{h}],\dots), \quad \mathbf{o}\in\mathcal{O}_j, \ j=0,\dots,x. \tag{1}$$

Indexing functions l_j and $r_{i,j}$ are affine mappings $l_j(\mathbf{o}) = \mathbf{L}_j\mathbf{o}+\mathbf{l}_j$ and $r_{i,j} = \mathbf{R}_{i,j}(\mathbf{o}+\mathbf{h})+\mathbf{r}_{i,j}$, respectively, with $\mathbf{L}_j \in \mathbb{Z}^{m_j\times n}$, $\mathbf{l}_j \in \mathbb{Z}^{m_j}$, $\mathbf{R}_{i,j} \in \mathbb{Z}^{q_{i,j}\times n}$, and $\mathbf{r}_{i,j} \in \mathbb{Z}^{q_{i,j}}$. Domains $\mathcal{O}_j$ and $\mathcal{H}_j$ are polyhedral subsets of $\mathbb{Z}$-modules $[\mathbf{B}^o]_j \subseteq \mathbb{Z}^n$ and $[\mathbf{B}^h]_j \subseteq \mathbb{Z}^n$, respectively. For affine indexed algorithms holds $[\mathbf{B}^h_j] \subseteq \mathcal{K}_j = \ker_{\mathbb{Z}^n}l_j$ and $[\mathbf{B}^o_j] \subseteq \mathcal{G}_j \cong \mathbb{Z}^n/\mathcal{K}_j$ [1] . Hence, all instances of $y_j[.]$ are uniquely defined. Φ is an associative and commutative operator such as $\sum, \prod, max$, and gcd.

Example 2.1 For the estimation of motion in a sequence of video frames, the current frame $\mathcal{C} = \{c[x,y], \ 0 \le x < S_h, \ 0 \le y < S_v\}$ of pixels $c[x,y]$ is subdivided into macro blocks $\mathcal{C}_{\tilde{x},\tilde{y}} = \{c[\tilde{x}+i,\tilde{y}+j], \ 0 \le i < M, \ 0 \le j < N\}$ of $M \times N$ pixels and the mean absolute error

$$\psi[\tilde{x},\tilde{y},\Delta x,\Delta y] := \frac{1}{MN} \sum_{i\in\{0,\dots,M-1\}} \sum_{j\in\{0,\dots,N-1\}} |c[\tilde{x}+i,\tilde{y}+j] - r[\tilde{x}+i+\Delta x,\tilde{y}+j+\Delta y]|,$$

---

[1]A/B denotes the quotient set.

$$0 \leq \tilde{x} < \frac{S_h}{M}, \ \tilde{x} \in M \cdot \mathbb{Z}, \ 0 \leq \tilde{y} < \frac{S_v}{N}, \ \tilde{y} \in N \cdot \mathbb{Z}, -\varepsilon \leq \Delta x, \Delta y \leq \varepsilon \quad (2)$$

with all blocks $\mathcal{R}_{\tilde{x},\tilde{y},\Delta x,\Delta y} = \{c[\tilde{x} + \Delta x + i, \tilde{y} + \Delta y + j], \ 0 \leq i < M, \ 0 \leq j < N\}$ for $-\varepsilon \leq \Delta x, \Delta y \leq \varepsilon$ of a past frame $\mathcal{R} = \{r[x,y], \ 0 \leq x < S_h, \ 0 \leq y < S_v\}$ is computed. The tuple $(\Delta x', \Delta y')$ for which $\psi[\tilde{x}, \tilde{y}, \Delta x, \Delta y]$ is minimum is denoted as motion vector [1]. The MAE-algorithm (2) is an affine indexed algorithm given by

$$\psi[l(\mathbf{o})] := \sum_{\mathbf{h} \in \mathcal{H}} |c[r_c(\mathbf{o} + \mathbf{h})] - r[r_r(\mathbf{o} + \mathbf{h})]|, \ \mathbf{o} \in \mathcal{O}.$$

Variables $\psi[.]$, $c[.]$, and $r[.]$ are affine indexed with indexing functions

$$\mathbf{L} = \begin{pmatrix} 1 & 0 & 0 & 0 & 0 & 0 \\ 0 & 1 & 0 & 0 & 0 & 0 \\ 0 & 0 & 1 & 0 & 0 & 0 \\ 0 & 0 & 0 & 1 & 0 & 0 \end{pmatrix}, \quad \mathbf{R}_c = \begin{pmatrix} 1 & 0 & 0 & 0 & 1 & 0 \\ 0 & 1 & 0 & 0 & 0 & 1 \end{pmatrix}, \quad \mathbf{R}_r = \begin{pmatrix} 1 & 0 & 1 & 0 & 1 & 0 \\ 0 & 1 & 0 & 1 & 0 & 1 \end{pmatrix}.$$

Domains of the indexing functions are

$$\text{dom } l = \mathcal{O} = \{\mathbf{o} = \mathbf{B}^o \mathbf{x}, \ \mathbf{A}^o \mathbf{o} \geq \mathbf{a}_0^o, \ \mathbf{x} \in \mathbb{Z}^4\},$$
$$\text{dom } r_c = \text{dom } r_r = \mathcal{O} \oplus \mathcal{H}, \quad \mathcal{H} = \{\mathbf{h} = \mathbf{B}^h \mathbf{x}, \ \mathbf{A}^h \mathbf{h} \geq \mathbf{a}_0^h, \ \mathbf{x} \in \mathbb{Z}^2\}$$

with

$$\mathbf{B}^o = \begin{pmatrix} M & 0 & 0 & 0 \\ 0 & N & 0 & 0 \\ 0 & 0 & 1 & 0 \\ 0 & 0 & 0 & 1 \\ 0 & 0 & 0 & 0 \\ 0 & 0 & 0 & 0 \end{pmatrix}, \quad \mathbf{A}^o = \begin{pmatrix} \mathbf{I}_4 \\ -\mathbf{I}_4 \end{pmatrix}, \quad \mathbf{B}_\psi^h = \begin{pmatrix} \mathbf{0}_{4 \times 2} \\ \mathbf{I}_2 \end{pmatrix}, \quad \mathbf{A}_\psi^h = \begin{pmatrix} \mathbf{I}_2 \\ -\mathbf{I}_2 \end{pmatrix}, \quad \mathbf{a}_0^h = \begin{pmatrix} 0 \\ 0 \\ 1-M \\ 1-N \end{pmatrix},$$

and $\mathbf{a}_0^o = \begin{pmatrix} 0 & 0 & -\varepsilon & -\varepsilon & M-S_h & N-S_v & -\varepsilon & -\varepsilon \end{pmatrix}^t$. $\mathbf{I}_m \in \mathbb{Z}^{m \times m}$ denotes the identity matrix and $\mathbf{0}_{m \times n} \in \mathbb{Z}^{m \times n}$ the null matrix. □

3 Embedding and localization

Data dependences of an affine indexed algorithm can be expressed by a digraph, the so-called data dependence graph. The vertex set $\mathcal{I} = \{\mathbf{i}\}$ is given by the union of all images $\mathcal{I}_l = \text{im } l$ and $\mathcal{I}_r = \text{im } r$. There is an edge from $\mathbf{i}_1$ to $\mathbf{i}_2$, if a $y_j[\mathbf{i}_2]$ is computed by a function with $y_i[\mathbf{i}_1]$ as argument.

First, an embedding of all images $\mathcal{I}$ in one and the same $\mathbb{Z}-$ module is applied. This allows an efficient analysis and modification of data dependences. The embedding leads to the indexing functions

$$l_j'(\mathbf{o}) = \mathbf{W}_j \mathbf{L}_j \mathbf{o} + \mathbf{w}_j, \quad r_{i,j}'(\mathbf{o} + \mathbf{h}) = \mathbf{W}_{i,j} \mathbf{R}_{i,j}(\mathbf{o} + \mathbf{h}) + \mathbf{w}_{i,j}.$$

For a localization of data dependences, (1) is subdivided into two statements

$$y_j[l'(\mathbf{i})] := \Phi a[\mathbf{i}], \quad \mathbf{i} \in \mathcal{I}, \quad (3)$$

$$a[\mathbf{i}] := f(y_i[r_{i,j}(\mathbf{i})], \dots]), \quad \mathbf{i} \in \mathcal{I} \quad (4)$$

with $\mathcal{I} = \mathcal{O} \oplus \mathcal{H}$. The subscripts j are omitted to the sake of brevity in the following. Let us consider statements (3) and (4) as relations $R(y_r[.], y_l[.])$ in $\{y_r[.]\} \times \{y_l[.]\}$, where $y_r[.]$ and $y_l[.]$ are variables on the right hand side and on the left hand side of a statement, respectively. For statement (3) holds

$$\forall a[\mathbf{i}] : \exists! y_j[l'(\mathbf{i})] : \ R(a[.], y_j[.])$$

and for statement (4) holds

$$\forall a[\mathbf{i}] : \exists! y_i \left[r'_{i,j}(\mathbf{i}) \right] : R(y_i[.], a[.])$$

We denote (3) as right-unique and (4) as left-unique statement. Vertices of a data dependence graph of (3) have an in-degree, which depends on $\mathcal{I}$ (see **Fig.**2b), and vertices of a data dependence graph of (4) have an out-degree, which depends on $\mathcal{I}$ (see **Fig.**2a). Furthermore, there may exist data dependences between vertices $\mathbf{i}_1$ and $\mathbf{i}_2$ which are not neighbors such that $-\mathbf{n} \leq \mathbf{i}_1 - \mathbf{i}_2 \leq \mathbf{n}$ holds. Object of localization is the generation of an algorithm coding such that all vertices of the corresponding data dependence graph have constant degree and such that there are only edges between neighboring vertices. This localization is done by replacing all data dependences by directed paths [15]. **Fig.**2 illustrates the localization of a left- and a right-unique statement. Solid edges represent

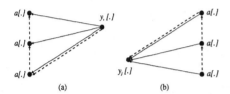 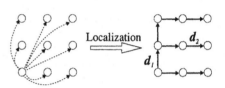

Figure 2: *Data dependences of* (a) *a left-unique and* (b) *a right-unique statement.*

Figure 3: *Localization of data dependences of a left-unique statement.*

the initially given data dependences and the dashed edges represent the directed paths introduced for localization.

Localization of data dependences from $\mathbf{i}$ to $\mathbf{WLi}$ in statement (3) is realized by a replacement of each direct data dependence by a directed path from $\mathbf{i}$ to $(\mathbf{I} - \mathbf{d}_1 \mathbf{b}_1^t)\mathbf{i}$, a path from $(\mathbf{I} - \mathbf{d}_1 \mathbf{b}_1^t)\mathbf{i}$ to $(\mathbf{I} - \mathbf{d}_1 \mathbf{b}_1^t - \mathbf{d}_2 \mathbf{b}_2^t)\mathbf{i}$, and so on with

$$\mathbf{W_L i} = \mathbf{i} - (\mathbf{d}_1 \mathbf{b}_1^t)\mathbf{i} - (\mathbf{d}_2 \mathbf{b}_2^t)\mathbf{i} - \cdots - (\mathbf{d}_m \mathbf{b}_m^t)\mathbf{i}.$$

In general, m is equal to the rank of $\mathbf{WL} - \mathbf{I}$. Furthermore, each edge from $\mathbf{R}_x \mathbf{i}$ to $(\mathbf{R}_x - \mathbf{d}_x \mathbf{b}_x^t)\mathbf{i}$ is replaced by a path of $\mathbf{b}_x^t \mathbf{i} + 1$ vertices $\{\mathbf{R}_x \mathbf{i} - \alpha \mathbf{d}_x, \ 0 \leq \alpha \leq \mathbf{b}_x^t \mathbf{i}\}$ [15]. Hence, for $m = 2$ statement (3) of the affine indexed algorithm is mapped onto the piecewise regular algorithm

$$
\begin{array}{lll}
z_1[\mathbf{i}] := & a[\mathbf{i}], & \mathbf{i} \in \mathcal{I}_1, \\
z_1[\mathbf{i}] := & z_1[\mathbf{i} - \mathbf{d}_2] + a[\mathbf{i}], & \mathbf{i} \in \mathcal{I}_2, \\
z_0[\mathbf{i}] := & z_1[\mathbf{i}], & \mathbf{i} \in \mathcal{I}_3, \\
z_0[\mathbf{i}] := & z_0[\mathbf{i} - \mathbf{d}_1] + z_1[\mathbf{i}], & \mathbf{i} \in \mathcal{I}_4, \\
\psi[\mathbf{i}] := & z_0[\mathbf{i}], & \mathbf{i} \in \mathcal{I}_5.
\end{array}
$$

Domains $\mathcal{I}_j = \{\mathbf{i} = \mathbf{B}\mathbf{x}, \ \mathbf{A}_j \mathbf{i} \geq \mathbf{a}_{0,j}\}$ are polyhedral subsets of a $\mathbb{Z}$−module $\mathcal{I}^\infty = [\mathbf{B}]$.

Example 3.1 Let us consider statement

$$y[\mathbf{i}] := f(x[0], \dots), \ \mathbf{i} \in \mathcal{I} = \{\mathbf{i} = \left(\begin{smallmatrix} x_1 \\ x_2 \end{smallmatrix} \right), \ 0 \leq x_j \leq 2\}$$

to illustrate a localization. We obtain from $\mathbf{d}_1 = \mathbf{b}_1^t = \left(\begin{smallmatrix} 0 \\ 1 \end{smallmatrix}\right)$, $\mathbf{d}_2 = \mathbf{b}_2^t = \left(\begin{smallmatrix} 1 \\ 0 \end{smallmatrix}\right)$

$$
\begin{aligned}
z_0[\mathbf{i}] &:= x[\mathbf{i}], & x_1 = x_2 = 0, \\
z_0[\mathbf{i}] &:= z_0[\mathbf{i} - \mathbf{d}_1], & x_1 = 0,\ 1 \le x_2 \le 2, \\
z_1[\mathbf{i}] &:= z_0[\mathbf{i}], & x_1 = 0,\ 0 \le x_2 \le 2, \\
z_1[\mathbf{i}] &:= z_1[\mathbf{i} - \mathbf{d}_2], & 1 \le x_1 \le 2,\ 0 \le x_2 \le 2, \\
y[\mathbf{i}] &:= f(z_1[\mathbf{i}], \dots), & 0 \le x_1 \le 2,\ 0 \le x_2 \le 2.
\end{aligned}
\tag{5}
$$

The data dependences of piecewise regular algorithm (5) are shown in **Fig.3**.   □

In piecewise regular algorithms, the indexing function of variables on the left hand side is the identity. Indexing function of variables on the right hand side are translations $\mathbf{i} - \mathbf{d}$, where $\mathbf{d}$ is denoted as data dependence vector.

4   Space-time-partitioning

In this section, we outline the determination of an initial scheduling and allocation of the vertices $\mathbf{i}$ of piecewise regular algorithms.

Module $\mathcal{I}^\infty$ is partitioned into to submodules $\mathcal{P}^\infty = [\mathbf{P}]$ and $\mathcal{S}^\infty = [\mathbf{S}]$ with

$$
\mathcal{I}^\infty = \mathcal{P}^\infty \oplus \mathcal{S}^\infty, \quad \mathcal{S}^\infty \cong \mathcal{I}^\infty / \mathcal{P}^\infty.
$$

Hence, for each vertex exists a uniquely defined representation

$$
\mathbf{i} = \mathbf{p} + \mathbf{s}, \quad \mathbf{p} \in \mathcal{P}^\infty, \ \mathbf{s} \in \mathcal{S}^\infty.
$$

Furthermore, we introduce a linear order in $\mathcal{S}^\infty$ with a relation $\sqsubset$

$$
\mathbf{s}_1 = \mathbf{S}\mathbf{z}_1 \sqsubset \mathbf{s}_2 = \mathbf{S}\mathbf{z}_2 \leftrightarrow \mathbf{z}_1 \prec \mathbf{z}_2,
$$

where $\mathbf{z}_1 \prec \mathbf{z}_2$ holds, if for $\boldsymbol{\delta} = \mathbf{z}_1 - \mathbf{z}_2$ $\delta_j > 0$ applies for the largest index j with $\delta_j \ne 0$. This scheduling satisfies the causality constraint imposed by the data dependences if $\forall \mathbf{d}_j = \mathbf{P}\mathbf{y}_j + \mathbf{S}\mathbf{z}_j : \ \mathbf{0} \prec \mathbf{z}_j$ applies. Here, we assume that all operations associated with a vertex $\mathbf{i}$ can be performed by a processor module in a time period $\Delta t = 1$. Furthermore, we use a scheduling function $\sigma(\mathbf{z}) = \boldsymbol{\tau}\mathbf{z} \in \mathbb{Z}$ with $\boldsymbol{\tau} \in \mathbb{Z}^{1 \times q}$, $q = \dim \mathcal{S}^\infty$, and

$$
\mathbf{z}_1 \prec \mathbf{z}_2 \leftrightarrow \sigma(\mathbf{z}_1) < \sigma(\mathbf{z}_2).
$$

If there are no resource constraints, then a vertex $\mathbf{i} = \mathbf{P}\mathbf{y} + \mathbf{S}\mathbf{z}$ is allocated to a processor module represented by $\mathbf{p} = \mathbf{P}\mathbf{y}$ and is computed at $t = \sigma(\mathbf{z})$. Hence, we obtain a set of processor modules

$$
\mathcal{P} = \{\mathbf{p}, \ \exists \mathbf{i} \in \mathcal{I} : \ \mathbf{i} = \mathbf{p} + \mathbf{s}, \ \mathbf{p} \in \mathcal{P}^\infty, \ \mathbf{s} \in \mathcal{S}^\infty\}
$$

for a data dependence graph with vertex set $\mathcal{I}$. Furthermore, an edge from $\mathbf{i}_1$ to $\mathbf{i}_2 = \mathbf{i}_1 + \mathbf{d}$ with $\mathbf{i}_j = \mathbf{P}\mathbf{y}_j + \mathbf{S}\mathbf{z}_j$ and $\mathbf{d} = \mathbf{P}\mathbf{y}_d + \mathbf{S}\mathbf{z}_d$ has to be realized by an interconnection $\mathbf{v} = \mathbf{P}\mathbf{y}_d$ from $\mathbf{P}\mathbf{y}_1$ to $\mathbf{P}\mathbf{y}_2$ with delay $\Delta t = \sigma(\mathbf{z}_d)$.

Fig.4a illustrates a space-time-partitioning of a vertex set $\mathcal{I} = \{\mathbf{i} = \left(\begin{smallmatrix} x_1 \\ x_2 \end{smallmatrix}\right),\ 0 \le x_j \le 3\}$ with $\mathbf{P} = \left(\begin{smallmatrix} 1 \\ 0 \end{smallmatrix}\right)$, $\mathbf{S} = \left(\begin{smallmatrix} 0 \\ 1 \end{smallmatrix}\right)$. Processor modules of the resulting array are represented by elements of $\mathcal{P} = \{\mathbf{p} = \mathbf{P}\mathbf{y},\ 0 \le y \le 3\}$. The data transfer induced by a data dependence vector $\mathbf{d} = \left(1\ 1\right)^t$ has to be realized by a interconnection $\mathbf{v} = \mathbf{P}$ with a time delay $\Delta t = 1$. In **Fig.4a**, interconnections are shown by edges between processor modules and associated memory elements are illustrated as black rectangles. Solid lines starting from processor modules symbolize the resulting scheduling and allocation.

5 Partitioning

The number $|\mathcal{P}|$ of processor modules resulting from a space-time partitioning depends on $\mathcal{I}$. The I/O-access to the peripheral memory is determined by the I/O-access of processor modules at the edge of the array. Hence, we have to provide an approach to scale the number of processor modules and the I/O-behavior of the array. Furthermore, the number of memory locations in the array depends on $\mathcal{I}$ for dim $\mathcal{S}^\infty > 1$, since $\Delta t = \sigma(\mathbf{z}_d)$ depends on $\mathcal{I}$ for dim $\mathcal{S}^\infty > 1$. The following example illustrates this dependence.

Example 5.1 We assume

$$\mathcal{S} = \{\mathbf{s}, \exists \mathbf{i} \in \mathcal{I} : \mathbf{i} = \mathbf{p} + \mathbf{s}, \ \mathbf{s} \in \mathcal{S}^\infty, \ \mathbf{p} \in \mathcal{P}^\infty\} = \{\mathbf{s} = \mathbf{S}\left(\begin{smallmatrix} z_1 \\ z_2 \end{smallmatrix}\right), \ 0 \le z_j < w_j\}.$$

Here, we obtain for example $\boldsymbol{\tau} = (\begin{smallmatrix} 1 & w_1 \end{smallmatrix})$. A data dependence vector $\mathbf{d} = \mathbf{Py} + \mathbf{S}\left(\begin{smallmatrix} 1 \\ 1 \end{smallmatrix}\right)$ has to be realized by an interconnection $\mathbf{Py}$ with time delay $\Delta t = 1 + w_1$, i.e. $1 + w_1$ memory locations have to be implemented, since w_1 new data are sent to this interconnection up to the consumption of a data by the receiving processor module.

□

Algorithm partitioning provides an approach to scale the number of processor modules, the I/O-behavior, and the number of memory locations of the array. There exist two different partitioning schemes, which differ with respect to the resulting scheduling and allocation of vertices $\mathbf{i}$ in the partitioned algorithm, e.g. [11, 12, 2]. In the locally sequential, globally parallel partitioning (LSGP), module $\mathcal{I}^\infty$ is partitioned into

$$\mathcal{I}^\infty = \mathcal{P}_\pi \oplus \mathcal{P}_\pi^\infty \oplus \mathcal{S}^\infty, \tag{6}$$

$$\cdot \mathcal{P}_\pi = \{\mathbf{i}_\pi = \mathbf{P}\boldsymbol{\kappa}, \ 0 \le \kappa_j < \vartheta_j\}, \ \mathcal{P}_\pi^\infty = [\mathbf{P}\boldsymbol{\Theta}^p]$$

with $\vartheta_j \in \mathbb{N}_+$ and $\boldsymbol{\Theta}^p = \mathrm{diag}(\vartheta_j)$. Subsets $\mathbf{p}_\pi^\infty + (\mathcal{P}_\pi \oplus \mathcal{S}^\infty)$ with $\mathbf{p}_\pi^\infty \in \mathcal{P}_\pi^\infty$ are denoted as partitions. All vertices of a partition are scheduled sequentially on a processor module represented by $\mathbf{p}_\pi^\infty$ and all partitions are active in parallel. **Fig.4b** illustrates an LSGP-partitioning of the data dependence graph shown in **Fig.4a**. Dashed rectangles symbolize partitions of $\mathcal{I}$. Solid lines starting from processor modules symbolize the resulting scheduling and allocation. This example shows that the number of memory locations in the array depends on $\mathcal{I}$ for a synthesis of an array with a defined number of processor modules. Furthermore, the time interval between accesses to the peripheral memory has been increased.

In the (standard) locally parallel, globally sequential partitioning (LPGS), module $\mathcal{I}^\infty$ is partitioned according to (6). We apply an extended version of this LPGS-scheme, where $\mathcal{I}^\infty$ is partitioned into

$$\mathcal{I}^\infty = \mathcal{I}_\pi \oplus \mathcal{I}_\pi^\infty, \mathcal{I}_\pi = \mathcal{P}_\pi \oplus \mathcal{S}_\pi, \ \mathcal{I}_\pi^\infty = \mathcal{P}_\pi^\infty \oplus \mathcal{S}_\pi^\infty,$$

$$\mathcal{I}_\pi = \{\mathbf{i}_\pi = \mathbf{B}\boldsymbol{\kappa}, \ 0 \le \kappa_j < \vartheta_j\}, \ \mathcal{I}_\pi^\infty = [\mathbf{P}\boldsymbol{\Theta}] \tag{7}$$

with $\vartheta_j \in \mathbb{N}_+$ and $\boldsymbol{\Theta} = \mathrm{diag}(\vartheta_j)$. Here, vertices of a partition $\mathbf{i}_\pi^\infty + \mathcal{I}_\pi$ are scheduled and allocated by an application of the above described space-time partitioning on $\mathcal{I}_\pi$ with a time offset for each partition. Hence, we have a locally parallel processing within a partition. All partitions are processed sequentially on the same array. The processor

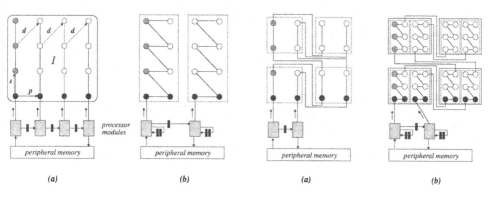

Figure 4: *Initial scheduling and alloca-tion* (a) *and scheduling and allocation by LSGP-partitioning* (b).

Figure 5: *Scheduling and allocation by LPGS-partitioning* (a) *and by co-partitioning* (b).

modules of this array are represented by $\mathcal{P}_\pi$. **Fig.**5a illustrates an LPGS-partitioning of the data dependence graph shown in **Fig.**4a. Solid lines starting from processor modules symbolize the resulting scheduling and allocation. This example shows that the number of memory locations in the array does not depend on $\mathcal{I}$ for a synthesis of an array with a defined number of processor modules. There is no increase of the time interval between accesses to the peripheral memory.

Co-partitioning [3, 4, 5] is a combined LPGS- and LSGP-partitioning. The number of pro-cessor modules is adjusted by the LPGS-partitioning and the number of memory locations and the access to the peripheral memory is adjusted by the LSGP-partitioning. **Fig.**5b illustrates a co-partitioning of the data dependence graph shown in **Fig.**4a. Dashed rect-angles symbolize LSGP-partitions and solid rectangles symbolize LPGS-partitions of $\mathcal{I}$. Solid lines starting from processor modules symbolize the resulting scheduling and allo-cation.

A co-partitioning introduces parameters ϑ_j in the algorithm code, which allow a scaling of parameters of the embedded array. A hierarchical co-partitioning introduces further parameters ϑ_j in the algorithm code, which allow a scaling of parameters of a peripheral memory hierarchy [5].

6 Relocalization

In this section, we consider the localization of data dependences of partitioned algorithms. **Fig.**6a shows a partitioned algorithm (5) with localized data dependences. A transfer of eight different variables between partitions and the peripheral memory has to be realized. However, all of these variables are copies of the same variable. Data transfers from/to the peripheral system are symbolized by fat drawn edges in **Fig.**6a. If partitions are scheduled sequentially as in the case of an LPGS-partitioning, then we can spent communication with and space on the peripheral memory by omitting localization between partitions as shown in **Fig.**6b. If tasks within a partition are scheduled sequentially as in the case of an LSGP-partitioning, then we can spent local memory allocated for each partition by omitting localization within partitions as shown in **Fig.**7a.

Relocalization provides an approach to a separat localization of data dependences

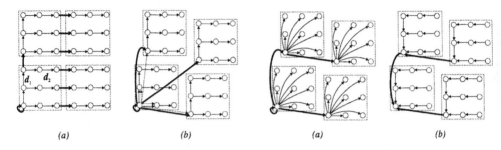

(a)   (b)   (a)   (b)

Figure 6:   (a) *Partitioned algorithm with localized data dependences,* (b) *LPGS-partitioned algorithm with relocalized data dependences*

Figure 7: (a) *LSGP-partitioned algorithm with relocalized left-unique statement,* (b) *LPGS-partitioned algorithm with relocalized right-unique statement*

within and between partitions. For this purpose, we introduce an embedding

$$e: \ \mathcal{I}^\infty = \mathcal{I}_\pi \oplus \mathcal{I}_\pi^\infty \longrightarrow \mathcal{I}^{*\infty} = \mathcal{I}_\pi \times \mathcal{I}_\pi^\infty$$

with

$$\mathcal{I}^\infty = [\mathbf{B}], \quad \mathcal{I}^{*\infty} = [\mathbf{B}^*] = \left[\begin{pmatrix} \mathbf{B} & 0 \\ 0 & \mathbf{B\Theta} \end{pmatrix} \right],$$

$\mathcal{I}^\infty \subseteq \mathbb{Z}^n$, $\mathcal{I}^{*\infty} \subseteq \mathbb{Z}^{2n}$ of the partitioned module $\mathcal{I}^\infty$ in module $\mathcal{I}^{*\infty}$. Space-time-, LPGS- and LSGP-partitioning in $\mathcal{I}^{*\infty}$ is given by

$$\mathcal{I}^{*\infty} = \mathcal{I}_\pi^* \oplus \mathcal{I}_\pi^{*\infty}, \quad \mathcal{I}_\pi^* = \mathcal{P}_\pi^* \oplus \mathcal{S}_\pi^*, \quad \mathcal{I}_\pi^{*\infty} = \mathcal{P}_\pi^{*\infty} \oplus \mathcal{S}_\pi^{*\infty}$$

with

$$\mathcal{P}_\pi^* = \left\{ \begin{pmatrix} \mathbf{P} \\ 0 \end{pmatrix} \kappa^p, \ 0 \le \kappa_j^p < \vartheta_j^p \right\}, \quad \mathcal{P}_\pi^{*\infty} = \left\{ \begin{pmatrix} 0 \\ \mathbf{P\Theta}^p \end{pmatrix} \kappa^{p,\infty}, \ \kappa^{p,\infty} \in \mathbb{Z}^{\dim \mathcal{P}^\infty} \right\},$$

$$\mathcal{S}_\pi^* = \left\{ \begin{pmatrix} \mathbf{S} \\ 0 \end{pmatrix} \kappa^s, \ 0 \le \kappa_j^s < \vartheta_j^s \right\}, \quad \mathcal{S}_\pi^{*\infty} = \left\{ \begin{pmatrix} 0 \\ \mathbf{S\Theta}^s \end{pmatrix} \kappa^{s,\infty}, \ \kappa^{s,\infty} \in \mathbb{Z}^{\dim \mathcal{S}^\infty} \right\}.$$

Processor modules of the array are represented by $\mathcal{P}_\pi^*$ in the case of an LPGS-partitioning and by $\mathcal{P}_\pi^{*\infty}$ in the case of an LSGP-partitioning. Furthermore, we transform statements (3) and statement (4) to

$$y_j \left[\begin{pmatrix} \mathbf{LB} & \mathbf{LB\Theta} \\ 0 & 0 \end{pmatrix} \mathbf{q} \right] := \Phi a \left[\begin{pmatrix} \mathbf{B} & 0 \\ 0 & \mathbf{B\Theta} \end{pmatrix} \mathbf{q} \right], \quad \mathbf{q} \in \mathcal{Q} \tag{8}$$

and

$$a \left[\begin{pmatrix} \mathbf{B} & 0 \\ 0 & \mathbf{B\Theta} \end{pmatrix} \mathbf{q} \right] := f \left(y_i \left[\begin{pmatrix} \mathbf{RB} & \mathbf{RB\Theta} \\ 0 & 0 \end{pmatrix} \mathbf{q} \right] \right), \quad \mathbf{q} \in \mathcal{Q} \tag{9}$$

with

$$Q = \left\{ q = \begin{pmatrix} \kappa \\ \kappa^{\infty} \end{pmatrix}, \ A^* q \ge a_0^* \right\}, \quad A^* = \begin{pmatrix} AB & AB\Theta \\ I & 0 \\ -I & 0 \end{pmatrix}, \quad a_0^* = \begin{pmatrix} a_0 \\ 0 \\ 1 - \vartheta \end{pmatrix}.$$

First, we consider the localization of the left-unique statement (9). We introduce the substitutions

$$a \left[\begin{pmatrix} B & 0 \\ 0 & B\Theta \end{pmatrix} q \right] := f \left(\mu \left[\begin{pmatrix} RB & 0 \\ 0 & B\Theta \end{pmatrix} q \right] \right), \quad q \in Q, \tag{10}$$

$$\mu \left[\begin{pmatrix} RB & 0 \\ 0 & B\Theta \end{pmatrix} q \right] := \eta \left[\begin{pmatrix} RB & 0 \\ 0 & RB\Theta \end{pmatrix} q \right], \quad q \in Q, \tag{11}$$

$$\eta \left[\begin{pmatrix} RB & 0 \\ 0 & RB\Theta \end{pmatrix} q \right] := y_i \left[\begin{pmatrix} RB & RB\Theta \\ 0 & 0 \end{pmatrix} q \right], \quad q \in Q, \tag{12}$$

where we restrict the notation to only one variable y_i on the right-hand side for the sake of brevity. Otherwise, we have to introduce a variable μ_i and η_i for each y_i. We denote the indexing function of a with l_a, of μ with r_μ, of η with r_η, and the indexing function of y_i with r_i. From ker $l_a \subseteq$ ker r_μ follows that statement (10) is left-unique. From $r_\mu(q) = 0$ follows $\kappa \in$ ker RB and $\kappa^{\infty} = 0$, since $B\Theta$ is the base of $\mathcal{I}_\pi^{\infty}$. We obtain ker $r_\mu \subseteq$ ker r_η. Hence, (11) is left-unique. From $r_\eta(q) = 0$ follows $\kappa \in$ ker RB and $\kappa^{\infty} \in$ ker RB. We obtain ker $r_\eta \subseteq$ ker r_i, i.e. (12) is left-unique.

If partitions are scheduled sequentially as in the case of an LPGS-partitioning, we localize only statement (10). Data dependences resulting from statements (11) and (12) are data dependences between sequentially processed partitions which result in communications between array and peripheral memory. Hence, only variables y_i are stored on the peripheral memory and have to be transferred to the data flow control unit between array and peripheral memory. Copies $\mu[.]$ and $\eta[.]$ of $y_i[.]$ are produced within the data flow control unit shown in **Fig.1**. This localization is shown in **Fig.6b**. Fat drawn edges symbolize the non-localized access to memory location $x[0]$ on the peripheral memory. Dashed lines symbolize the four dimensions of $\mathcal{I}^*$. In general, the array resulting from this relocalization may contain more processor modules than the array resulting from partitioning, since $r_i(\mathcal{I}_\pi) \not\subseteq \mathcal{I}_\pi$ may hold or since there may exist edges of localization paths outside the original partition $\mathcal{I}_\pi$. Additional processor modules are only memory cells for a realization of necessary routing operations.

Now we consider relocalization if partitions $i^{\infty} + \mathcal{I}_\pi$ are processed in parallel and the operations within a partition are processed sequentially as in the case of an LSGP-partitioning. Here, we localize only statement (11), since only data dependences resulting from this statement induce communications between processor modules of the array architecture. Data dependences of the left-unique statements (10) and (12) remain non-localized. This localization is shown in **Fig.7a**. The gain of this localization is a decrease of the allocated memory capacity within the array. Only variables y_i are stored on the peripheral memory and have to be transferred to the data flow control unit. Copies $\eta[.]$ of $y_i[.]$ used in the

array are produced within the data flow control unit shown in **Fig.**1.
For the relocalization of the right-unique statement (8), we introduce substitutions

$$y_j \left[\begin{pmatrix} \text{LB} & \text{LB}\Theta \\ 0 & 0 \end{pmatrix} q \right] := \Phi_\xi \left[\begin{pmatrix} \text{LB} & 0 \\ 0 & \text{LB}\Theta \end{pmatrix} q \right], \quad q \in \mathcal{Q}, \tag{13}$$

$$\xi \left[\begin{pmatrix} \text{LB} & 0 \\ 0 & \text{LB}\Theta \end{pmatrix} q \right] := \Phi_\rho \left[\begin{pmatrix} \text{LB} & 0 \\ 0 & \text{B}\Theta \end{pmatrix} q \right], \quad q \in \mathcal{Q}, \tag{14}$$

$$\rho \left[\begin{pmatrix} \text{LB} & 0 \\ 0 & \text{B}\Theta \end{pmatrix} q \right] := \Phi_a \left[\begin{pmatrix} \text{B} & 0 \\ 0 & \text{B}\Theta \end{pmatrix} q \right], \quad q \in \mathcal{Q} \tag{15}$$

Statements (13)-(15) are right-unique. Relocalization for right-unique statements can be done in a similar way as for left-unique statements. However, for LPGS-partitioning statements (14) and (15) have to be localized too. **Fig.**7b illustrates the relocalization of a right-unique statement for an LPGS-partitioning. If we restrict the application to cases, where $l(\mathcal{I}_\pi) \subseteq \mathcal{I}_\pi$ holds, then (13) is left- and right-unique, since $l(\mathcal{I}_\pi^\infty) \subseteq \mathcal{I}_\pi^\infty$ and $\mathcal{I}_\pi \cap \mathcal{I}_\pi^\infty = \{0\}$ apply. Hence, we can merge (13) and (14) after a reindexing of y_j to

$$y_j \left[\begin{pmatrix} \text{LB} & 0 \\ 0 & \text{LB}\Theta \end{pmatrix} q \right] := \Phi_\rho \left[\begin{pmatrix} \text{LB} & 0 \\ 0 & \text{B}\Theta \end{pmatrix} q \right], \quad q \in \mathcal{Q}. \tag{16}$$

Finally, for LSGP-partitioning statement (13) remains non localized.

7   Conclusion

System synthesis from completely localized code leads to very inefficient utilization of the memory system, to high implementation and communication costs, and hence, to high power consumption. We proposed an approach to localize data dependences of partitioned algorithms which fits into the existing framework of array synthesis and overcomes these problems. The real gain of this approach is unknown until the effort of the resulting address computation has been evaluated.

Especially for real time image processing algorithms array architectures are well suited. These algorithms contain variables with indexing functions r with $\dim \ker r \geq 2$. In this case, relocalization is mandatory if partitioning has to be applied.

References

[1] V. Bashkaran and K. Konstantinides. *Image and Video Compression Standards.* Kluwer Academic Publishers, second edition, 1997.

[2] J. Bu and E. Deprettere. Processor clustering for the design of optimal fixed-size systolic arrays. In E. Deprettere and A.-J. van der Veen, editors, *Algorithms and Parallel VLSI Architectures*, volume A:Tutorials, pages 314–362. Elsevier Science Publisher B.V., 1991.

[3] U. Eckhardt and R. Merker. Co-partitioning - a method for hardware/software code-sign for scalable systolic arrays. In R. Hartenstein and V. Prasanna, editors, *Reconfigurable Architectures*, pages 131–138. IT Press, Chicago,IL, 1997.

[4] U. Eckhardt and R. Merker. Scheduling in copartitioned array architectures. *Proc. Int. Conf. Application-Specific Systems, Architectures, and Processors (ASAP)*, pages 219–228, 1997.

[5] U. Eckhardt and R. Merker. Hierarchical algorithm partitioning at system level for an improved utilization of memory structures. *IEEE Trans. on Computer-Aided Design*, 18(1):14–24, January 1999.

[6] R.M. Karp, R.E. Miller, and S. Winograd. The organisation of computations for uniform recurrence equations. *Journal of the Association of Computing Machinery*, 14(3):563–590, July 1967.

[7] H.T. Kung and C.E. Leiserson. Algorithms for VLSI processor arrays. In C. Mead and L. Conway, editors, *Introduction to VLSI systems*, chapter 8.3, pages 271–292. Addison-Wesley Publishing Company, 1980.

[8] E.J. McCluskey. Iterative combinational switching networks-general design consideration. *IRE trans. on electronic computers*, EC-7:258–291, 1958.

[9] W.L. Miranker and A. Winkler. Spacetime representations of computational structures. *Computing*, 32:93–114, 1984.

[10] D.I. Moldovan. On the ananlysis and synthesis of VLSI algorithms. *IEEE Trans. on Computers*, C-31(11):1121–1126, November 1982.

[11] D.I. Moldovan and J.A.B. Fortes. Partitioning and mapping algorithms into fixed size systolic arrays. *IEEE Trans. on Computers*, C-35(1):1–12, January 1986.

[12] J.J. Navarro, J.M. Llaberia, and M. Valero. Partitioning: An essential step in mapping algorithms into systolic array processors. *IEEE Computers*, pages 77–89, 1987.

[13] J.v. Neumann. The theory of self-reproducing automata. In A.W. Burks, editor, *Essays on cellular automata*. University of Illinois Press, Urbana, 1970.

[14] S.K. Rao. *Regular Iterative Algorithms and their Implementation on Processor Arrays*. PhD thesis, Information Systems Lab., Stanford University, October 1985.

[15] V.P. Roychowdhury, L. Thiele, S.K. Rao, and T. Kailath. Design of local VLSI processor arrays. In *Proc. Int. Conf. on VLSI and Signal Processing*, Monterrey, November 1988.

[16] V.P. Roychowdhury, L. Thiele, S.K. Rao, and T. Kailath. On the localization of algorithms for VLSI processor arrays. In *VLSI Signal Processing III*, pages 459–470, New York, 1989. IEEE Press.

[17] L. Thiele. Compiler techniques for massive parallel architectures. In P. Dewilde and J. Vandewalle, editors, *Computer Systems and Software Engineering*, pages 101–150. Kluwer Academic Publishers, 1992.

High-level condition expression transformations for design exploration

Martin Palkovic[‡]   Miguel Miranda   Francky Catthoor[†]   Diederik Verkest
IMEC Lab., Kapeldreef 75, 3001 Leuven, Belgium.
[‡]Department of Microelectronics, Faculty of Electrical Engineering and
 Information Technology, Slovak Univ. of Technology, Bratislava, Slovakia.
[†]Also Professor at Katholieke Univ. Leuven.

Abstract

Data intensive applications (i.e., multimedia) are clearly dominated by data transfer and storage issues. However, after removing the data transfer and address related bottlenecks, the control-flow mapping issues remain as important implementation overhead in a custom hardware realisation. The source of this overhead can be due to the presence of complex conditional code execution, loops or the mixed of both. In this work, we focus on optimising the behaviour of the conditional code which is dominated by complex condition test expressions. Our transformations aim in a first stage at increasing the degree of mutually exclusiveness of the initial condition trees. This step is complemented by optimising the decoding of the test expressions. In a second stage, architecture exploration is performed by trading-off at the high-level gate count against critical-path delay for the resulting code. We demonstrate the proposed transformations on a real-life driver using conventional behavioral synthesis tools as synthesis back-end. The driver selected represents the crucial timing bottleneck in a scalable architecture for MPEG-4 Wavelet Quantisation. Using our approach, we have explored in a very short time the design space at the high level and we have obtained a factor 2 reduction of the critical path with a smaller gate count overhead when compared to traditional RT or high-level synthesis based approaches, even when applied by experienced designers

1   Introduction

In conventional design flows, designers directly make one big jump from the system specification to a register transfer level description without the possibility of any significant exploration of the design alternatives at the higher levels. By making such a big jump, designers hope to save design time, but they cannot evaluate the impact of all the decisions they are implicitly making. This lack of extensive high-level exploration pushes most system-level problems down to solving any timing or cost related problem at the time-consuming logic synthesis phase. This significantly increases the search space at that level and hence, the CPU time required to perform the bit-level optimisations. This usually ends-up into unaffordable design iteration loops that usually cause a delay in the product delivery. This is true for both data and control-flow dominated applications

We propose a high-level exploration and synthesis methodology for control-flow related issues where bottlenecks are exposed much earlier in the design strategy than with research based or commercial High-Level Synthesis (HLS) tools with a scheduling and allocation step. This is done by spending more time than industrial designers usually do at the high level by using a

R. Merker and W. Schwarz (eds.), System Design Automation, 56-64.
© 2001 *Kluwer Academic Publishers.*

systematic exploration methodology which identifies promising paths in the search space and which removes uninteresting or invalid paths. When the designer is sure that the version obtained so far will meet the timing constraints, he can focus on reducing the main cost. This avoids the time-consuming RT-synthesis based design iterations, hence allowing fast and accurate architecture exploration and shorter design iteration cycles. In this way, the overall required time to go from specification to gates is actually shorter than the required time using the conventional approach.

To efficiently reuse existing HLS based design flows [1] and existing tool support [2] we propose to schedule the application using a library of application-specific resources instead of using a more traditional HLS approach. These resources are assigned to the complex code as targets during the behavioral synthesis stage. Our synthesis approach not only differentiates in performing the allocation and assignment at a different granularity level but also by performing this resource allocation and assignment steps prior to scheduling.

Most related work in optimising transformations and synthesis techniques for HLS of control-flow dominated applications have mainly targeted lower levels of abstraction, e.g. instruction level. These transformations are typically performed right before [3, 4, 5] or during the scheduling phase [6, 7] in HLS. In this paper, we propose to explore the design space for control-flow at a higher abstraction level, e.g., the condition (expression) level, and to efficiently bridge the gap between this (expression) level and the instruction level at which traditional HLS design methodologies are placed. This condition expression focus is also the main focus of our paper as opposed to our earlier CATHEDRAL-3/ADOPT work oriented to arithmetic dominated applications and especially address expressions [8, 9, 10]. Also, compilers do not perform these transformations as they are focussing on a programmable processor with different opportunities for instruction level parallelism as opposed to custom hardware.

We have applied our transformational approach for (re)designing the arithmetic coder subsystem of a wavelet based image compression engine. Feedback gathered with previous design experience [11, 12] have identified the selected driver as the timing bottleneck in the complete system chain [12]. Using our approach we have been able to explore design alternatives in much less time than required when a more conventional RT-based design flow is applied [13]. Moreover, we have been able to improve the performance with a factor of 2 and design productivity with a factor of 2-3 at an equivalent implementation (gate-count) cost.

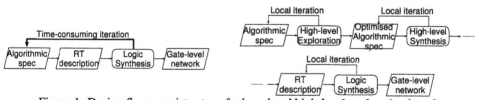

Figure 1: Design flows: register-transfer based and high-level exploration based.

2   Test driver description

The test driver chosen (see Figure 2) is part of an MPEG-4 wavelet compression engine. There are two main outputs in the compression engine: an image composed of average values (DC image) and a set of images with additional information that incrementally increases the quality of the DC image. During the decoding phase, the average image has to be processed entirely, unlike the additional detailed information which can be decoded according to the available performance of the target implementation platform. The arithmetic coder is part of a hardware accelerator block called the OZONE [12] chip. The OZONE chip performs also a zero-tree coding compression algorithm in the detailed images, up-front to the actual arithmetic coding algorithm.

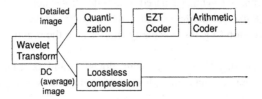

Figure 2: MPEG Wavelet image compression

An algorithmic description of the arithmetic coder [14] has been selected as test-driver. A C-language description of the driver is available from previous "in-house" design experiences. This driver is heavily dominated by conditional constructs and by a large amount of linear arithmetic computations. The code was obtained by a design team at IMEC as starting point for the generation of a RT-description. Therefore, it incorporates some of the original design refinement decisions (i.e., detailed data-types, distribution of the available cycle budget available across loops, etc). To make a fair comparison of results after synthesis, we have decided to keep these implementation decisions as starting point for our exploration phase and to explore only the remaining search space available.

From this code, a RT-description has been derived and synthesised at the logic level by our original design team. This has required two long design iterations in order to obtain an implementation that could be successfully synthesised at the logic level. The design iterations were mainly invested in functionally partitioning the description into two pieces so each piece could be successfully synthesised and (re)combined later.

```
// Original behavioral code in critical loop body

if(!(((!cc0&&!model)||(!cc1&&model))&&!model) {
    b1 = ...
    flag = ...
    b2 = ...
}
if(!(((!cc0&&!model)||(!cc1&&model))&&model) {
    b1 = ...
    flag = ...
    b2 = ...
}
if(((!cc0&&!model)||(!cc1&&model))&&!model) {
    b1 = ...
    flag = ...
    b2 = ...
}
if(((!cc0&&!model)||(!cc1&&model))&&model) {
    b1 = ...
    flag = ...
    b2 = ...
}
```

Figure 3: Illustration of mutually exclusive trafos: original behavioral code for conditionals inside time-critical loop

3   Design exploration and optimisation of the arithmetic coder

Our transformational based design flow is divided into two main stages. The first stage aims at obtaining an optimised description of the application at the same abstraction level as the initial one. This is done without having to trade-off e.g., throughput (cycle-count and/or delay) and cost (gate count). In this first stage, we focus on two issues: improving the degree of mutual exclusiveness in nested condition trees, and in optimising the decoding of the condition testing in deeply nested conditional constructs. In a second stage, architecture exploration is performed by focussing on meeting the timing constraints. This is applied to the control-flow by accelerating the evaluation of 'cycle' critical conditionals when mapping them into data-path multiplexing logic.

Both stages are complemented with arithmetic oriented transformations similarly to what is proposed in our ADOPT script [10] (also divided in equivalent stages with equivalent aims). The goal is to arrive to an optimised description of the application that can be efficiently synthesised using conventional HLS back-end tools.

3.1 High-level transformations for exclusive condition expressions

Typically, every conditional evaluation code introduces controller states and cycles during the scheduling phase, irrespective of the compiler and/or synthesis tool used. Some schedulers [6, 7] perform optimisations in order to reduce the required number of controller states. However, the abstraction-level at which those schedulers currently work is too low. Hence the search space is too large to efficiently arrive to a minimal required number of controller states.

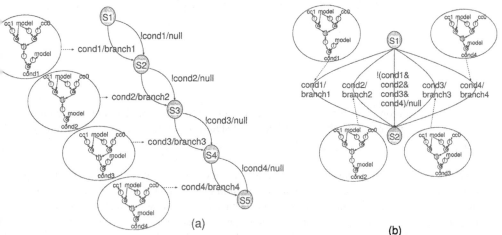

Figure 4: Illustration of mutually exclusive trafos: (a) initial state-transition diagram (STD) with clustered condition expressions; (b) STD after mutually exclusive analysis where false transitions (hence states) have been removed as a result of the mutually exclusion analysis. The clusters of condition expressions are illustrated using data-flow semantics. The transitions between states are executed after the conditions are evaluated. When the conditions are true, the corresponding branch (indicated by the transition edge) is executed, otherwise no behaviour is executed (null).

A large part of the search space can be efficiently pruned by exploring the opportunities for controller states (namely csteps) sharing when the code exposes explicitly the mutual exclusiveness of the conditional branches to the underlying scheduler. This mutual exclusiveness is in many cases inherent to the original specification code but not properly expressed by the application designer. Figure 3 shows an original piece of code of the arithmetic coder.

Our methodology starts by clustering the condition test expression graphs of the application by using a Control and Data-Flow (CDFG) model (see Figure 4a). This is done by maximally collecting the data-flow behaviour associated to each condition test expression. To maximally collect the data-flow behaviour of each condition expression we back-trace from the edge controlling the condition test backwards in the CDFG until another control-flow boundary, functional input or memory access operation is found. The degree of mutually exclusiveness of the resulting test expression graphs can be analysed by using standard bit-level graph theory techniques (e.g., binary decision diagrams or alike). After this analysis, the false control-flow dependencies are identified and removed from the graph (see Figure 4b).

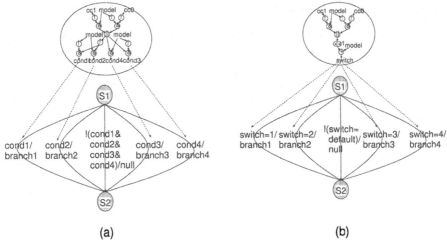

(a)   (b)

Figure 5: Illustration of mutually exclusive trafos: (a) state-transition diagram after condition expression merging using algebraic properties; (b) state-transition diagram after condition encoding.

At this point, further optimisation in the logic operation cost can be done by exploiting algebraic transformations [17] across condition expression graphs and by deriving a merged version where word and bit-level common subexpression have been identified and eliminated (see Figure 5a). The resulting merged condition expression graph is a one-hot decoded version of the four conditional branches. To efficiently express the mutually exclusive semantics of our transformation at the code level, we need to encode the four independent condition test expression into just one (see Figure 5b). Conventional techniques can be used [16] to encode the one-hot decode version of the four conditional branches.

Finally, the switch construct commonly found in C or HDL languages can be used to express the mutually exclusive semantics (see Figure 6) and the resulting expression can be encapsulated onto an ASU to avoid the traditional operation-level time-sharing overhad [8].

3.2   High-level time-multiplexing exploration of condition expressions

In a second step, and in order to efficiently reuse HLS as synthesis back-end [2], we encapsulate the resulting (merged) conditional graphs into Application Specific Units [8] (ASUs) where the functionality is then executed within one controller state. These ASUs are used as predefined resources during scheduling. In the design of the arithmetic coder, we have encapsulated each (merged) condition expression onto a different ASU.

```
// Optimised behavioral code in          // Optimised description of ASU for
// critical loop body                    // condition expression evaluation

switch (asu_switch(cc0,cc1,model)) {     asu_switch (int cc0, int cc1,
  case 0 :                                           int model) {
    b1 = ...
    flag = ...                               return (((!cc0&&!model)||
    b2 = ...                                          (!cc1&&model)<<1) + model);
  break;
  case 1 :                               }
    b1 = ...
    flag = ...
    b2 = ...
  break;
  case 2 :
    b1 = ...
    flag = ...
    b2 = ...
  break;
  case 3 :
    b1 = ...
    flag = ...
    b2 = ...
  break;
}
```

Figure 6: Illustration of mutually exclusive trafos: resulting behavioral code with RT-level code for ASUs.

This implementation, also referred as the hardwired implementation, provides the best timing implementation since no time-multiplexing opportunities are left out to the scheduler. After obtaining a hardwired implementation, a more effective architecture exploration can be performed by time-multiplexing different (merged) condition clusters into the same ASU. Only similar [9] (non-overlapping) optimised condition expressions are allowed to share the same ASU to minimise the time-multiplexing overhead. Still, regularity improvement amongst (merged/shared) clusters can be done by exploiting algebraic transformations at the local (ASU-level) scope [15]. In this way, the designer can fully trade-off timing (critical-path) against implementation cost (gate-count). Early feedback for both timing and cost related issues is possible by iterating after the scheduling phase of HLS, and before the time-consuming logic synthesis phase takes place.

Not only condition expression clusters can be encapsulated onto ASUs, but also the whole conditional construct (including the multiplexing decision and branch data-flow graphs) can be accelerated in the data path without being actually mapped onto controller states. In this way, we avoid the introduction of unnecessary csteps in the body of time-critical loops by the scheduler. However, these situations may lead to an increase of the size of the critical-path delay. A very efficient way of exposing such transformations prior to our HLS back-end [2] is to encapsulate the whole conditional construct into an ASU as similarly done for the condition test expressions. We will then use these as resources during the scheduling. Figure 7a shows a piece of the MPEG4 wavelet filter code, also presente in the critical loop, where this transformation has been applied to obtain an equivalent behaviour. Figure 8b) shows how one state has been removed from the global state transition diagram. Figure 7b shows the resulting behavioral code and the RT-level description of the ASUs.

In principle, this type of transformations should be performed by state-of-the-art schedulers. However, in practice commercially available schedulers [2] do not offer such optimisation degree. Therefore, we need to explore these alternatives before HLS and expose it to the underlying synthesis back-end.

Our control-flow time-multiplexing strategy consists of encapsulating non-related nested con-

```
// Original behavioral code        // Optimised behavioral code
// in critical loop body           // in critical loop body

if(a == 1) {                       asu_foo1 (a,count);
  count++;                         ...
  a = 8;                           if(a==1)
}                                    goo(a, count);
else                               else
  a -= 1;                            asu_foo2(a,count);
...
if(a == 1) {                       // Optimised description of ASUs for condition
  count++;                         // tree evaluation
  a = 7;
}                                  asu_foo1(int& a,          asu_foo2(int& a,
else {                                      int& count){               int& count){
  if(a == 2) {                       if(a == 1) {              if(a == 2) {
    count++;                           count++;                  count++;
    a = 8;                             a = 8;                    a = 8;
  }                                  }                         }
  else                               else {                    else {
    a -= 2;                            count = count;            count = count;
}                                      a -= 1;                   a -= 2;
                                     }                         }
                                   }                         }
```

(a)   (b)

Figure 7: Implementing control-flow as data-flow: (a) original behavioural code; (b) resulting behavioral code with RT-level code for ASUs.

dition trees into clusters of behaviour. The amount of control-flow that has been mapped onto ASUs has been steered by the budget available in each loop of our application. We transform the control-flow behaviour as data-flow [8] so as to make sure that the scheduler won't allocate more cstates than required by the available budget. Later, to reduce gate-count overhead we can trade-off timing (delay in critical-path) by time-multiplexing clusters of "similar" non-overlapping control-flow behaviour into the same ASU.

4   Results

We have performed synthesis experiments[1] to obtain results when applying HLS directly from the initial code and by applying our high-level transformation stage up-front and using a hardwired architecture mapping approach (see section 3.2). We have also compared the results obtained by our design team when using traditional RT-based design flows [12]. Table 1 shows the results obtained in gate count and delay for all three implementations after logic synthesis (by enabling maximum optimisation options). Our implementation approach shown in Section 3.1 leads to 7% improved gate-count overhead when compared to the conventional RT-based approach and 20% better when compared to traditional HLS. In addition, the critical path in our case gets improved with more than a factor of 2. This would allow to reduce the gate count even further by exploiting the multiplexing freedom. Design time has also been dramatically improved with a factor 2-3 and this with an inexperienced designer including the training. This demonstrates the power of the systematic methodology.

We have also performed architecture exploration by trading off timing and gate-count at the high-level using our condition expression sharing transformations shown in Section 3.2. Table 2 shows the gate count after logic synthesis (by enabling medium effort optimisation options), the critical-path in nanoseconds obtained after HLS (Delay estim.) and the critical path obtained

---

[1]Using a CMOS .5um standard cell libraryas target library.

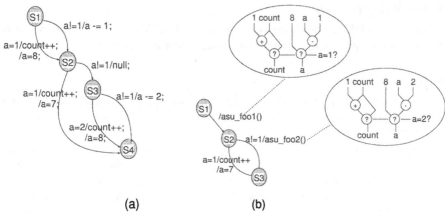

(a)   (b)

Figure 8: Implementing control-flow as data-flow: (a) initial state transition diagram; (b) state transition diagram after time multiplexing transformations.

after logic synthesis (Delay actual). The starting architecture (Init.+Optim.) is based on a fully hardwired implementation of the arithmetic coder after the application of our optimising transformation techniques. The other two architectures (Light_share & Heavy_share) are obtained by sharing condition expression clusters into a progressively smaller number of ASUs. We observe a significant saving in gate count at the expenses of a small delay overhead when comparing aggressive against moderate sharing. This proves the need for a systematic aggressive sharing methodology.

	RT	HLS	Trafo+HLS
Gate-count	15500	17508	14485
Delay (ns)	31.25	39.39	13.79

Table 1: Comparison of different synthesis approaches (compile -map_effort high)

Design	Init+Optim	Light_share	Heavy_share
Gate-count	18217.5	18206.2	11002.0
Delay estimations (ns)	40	50	70
Delay actual (ns)	17.43	29.40	32.27

Table 2: Architecture exploration via condition expression sharing (compile -map_effort low) - synthesis results

5 Conclusions

In this paper we propose a set of high-level transformation techniques for applications dominated by complex condition expressions. Our optimisation approach is based on exploiting global scope optimisation opportunities at the condition expression level instead of the lower instruction level. We have verified this using a real-life test driver subsystem part of a wavelet base image compression chain. Using our approach, we have explored in a very short time the design space at the high level and we have obtained a factor 2 reduction of the critical path with a smaller gate count overhead when compared to traditional RT or high-level synthesis based approaches applied by an experienced designer.

Acknowledgements

The authors gratefully acknowledge our colleagues at IMEC Mercedes Peon, Bart Vanhoof and Jan Bormans for providing the application driver code and support in the experiments.

References

[1] M.McFarland, A.Parker, R.Camposano, *The high-level synthesis of digital systems*, Proc. of the IEEE, special issue on "The future of computer-aided design", Vol. 78, No. 2, pp. 301-318, Feb. 1990.

[2] D.Knapp, T.Ly, D.MacMillen, R.Miller, *Behavioral Synthesis Methodology for HDL-Based Specification and Validation*, Proc. of 32nd ACM/IEEE Design Automation Conf., 1995

[3] P.Duzy, H.Krämer, M.Neher, M.Pilsl, W.Rosenstiel, T.Wecker, "CALLAS: Conversion of algorithms to library adaptable structures". Proc. VLSI Signal Procesing Workshop II, pp. 197-209, 19889.

[4] R.Camposano, "Behaviour preserving transformations for high-level synthesis", In Proc. Workshop on Hardware Specification, Verification and Synthesis, 1989.

[5] V.Chaiyakul, D.Gajski, L.Ramachandran, "High-level transformations for minimising syntatic variance", Proc. Workshop on High-Level Synthesis, 1992.

[6] W.Wolf, A.Takach, T-C.Lee, "Architectural optimization methods for control-dominated machines", in *Trends in high-level synthesis*, R.Camposano, W.Wolf (eds.), Kluwer Acad. Publ., Boston, 1991.

[7] A.Jerraya, *et. al.*, "System-level modelling for synthesis", Proc. Euro-DAC, 1995.

[8] S.Note, W.Geurts, F.Catthoor, and H.De Man. Cathedral-III: Architecture driven high-level synthesis for high throughput dsp applications. In *Proc. 28th ACM/IEEE Design Automation Conf.*, pages 597–602, June 1991.

[9] W.Geurts, F.Catthoor, and H.De Man. Time constrained allocation and assignment techniques for high throughput signal processing. In *Proc. 29th ACM/IEEE Design Automation Conf.*, pages 124–127, June 1993.

[10] M.Miranda, F.Catthoor, M. Janssen, H.De Man, *High-Level Address Optimization and Synthesis Techniques for Data-Transfer Intensive Applications*, IEEE Trans. on VLSI Systems, no.4, vol.6, pp. 677-686. December, 1998.

[11] M. Peon, G. Lafruit, B. Vanhoof, and Jan Bormans, *Design of an Arithmetic Coder for a Hardware Wavelet Compression Engine*, SPS-98, Belgium, March 1998.

[12] B.Vanhoof, M.Peon, G.Lafruit, J,Bormans, L.Nachtergaele, and I.Bolsens, *A Scalable Architecture for MPEG-4 Wavelet Quantization*, Journal of VLSI Signal Processing Systems for Signal, Image, and Video Technology, Vol. 23, No. 1, pp. 93-107, October 1999.

[13] S.Carlson, *Introduction to HDL-based design using VHDL*, Synopsys Inc, California, 1990.

[14] G.Held, T.R.Marshall, *Data and image compression: Tools and Techniques*, John Wiley & sons, 1996.

[15] J.M.Janssen, F.Catthoor, H.De Man, *A specification invariant technique for regularity improvement between flow-graph clusters*, Proc. 7th ACM/IEEE European Design & Test Conf., pp. 138-143, 1996.

[16] G.De Micheli, "Synthesis and optimisation of digitial circuits", McGraw-Hill, NJ, 1994.

[17] J.M.Janssen, F.Catthoor, and H.De Man. A specification invariant technique for operation cost minimization in flow-graphs. In *Proc. 7th ACM/IEEE International Symposium on High-level Synthesis*, pages 146–157, 1994.

Component-based System Design in Microkernel-based Systems

Lars Reuther*   Volkmar Uhlig†   Ronald Aigner*

*Dresden University of Technology
Institute for System Architecture
{reuther,ra3}@os.inf.tu-dresden.de

†University of Karlsruhe
Institute for Operating- and
Dialoguesystems
volkmar@ira.uka.de

Abstract

Component-based software design is a widely accepted approach to deal with the growing demands of current software systems. Existing component models are targeted towards flexible software design and load distribution between multiple nodes. These systems are mainly designed for interoperability. Thus, they are very general and flexible, but slow. Building a microkernel-based system using existing component technology would result in bad overall system performance. We propose an approach to overcome the limitations of existing component systems while maintaining their advantages. This paper gives an overview of a new IDL compiler, *FIDL*, which uses knowledge of the underlying communication mechanism to improve the performance of component-based systems.

1 Introduction

Microkernel-based systems are gaining more and more attention. They provide a flexible approach to deal with the complexity of operating systems by dividing systems into smaller units or *components*. However, early systems like Chorus [10] or Mach [4] suffered from poor inter-process communication (IPC) performance. This resulted in the common opinion that microkernel-based system are inherently slow. Recent work [7, 6] has shown that modern microkernel architecture can improve IPC performance significantly and that microkernel-based systems can reach the performance of traditional monolithic systems.

However, there is another problem such systems have to attack: usability. The microkernel provides general abstractions such as address-space protection and IPC. While this enables the kernel developer to highly optimize the microkernel, it is difficult to build large systems with these low-level abstractions. Building a complex communication interface using only the microkernel interface is time consuming and error-prone. Instead, a software designer needs ways to specify the component interfaces at a higher level. Invocation-code can be generated automatically from these specifications.

A closer look at the structure of microkernel-based systems shows that they are quite similar to *Distributed Systems* like *CORBA* [9] or *DCOM* [2]. Both types of systems are designed of

R. Merker and W. Schwarz (eds.), System Design Automation, 65-72.

several servers—or components—which interact. In distributed systems, multiple components interact via well defined interfaces. These are declared in special languages, *Interface Descrip-tion Languages (IDL)*. *IDLs* are languages that describe interfaces between components. They are independent of the programming language which is used to implement the components. An IDL compiler generates the function stubs for both the sender (client) and receiver (server) of an IPC (Fig. 1).

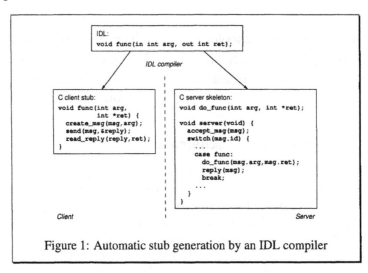

Figure 1: Automatic stub generation by an IDL compiler

The generated stubs perform two operations:

1. Convert the arguments of the function call to/from the message buffer (*marshaling*).

2. Do the IPC call.

The need for Interface Description Languages in Distributed Systems arose for various reasons:

- Support for software design,

- Automatic code generation, eliminating common sources of errors in the software devel-opment.

- Code reuse and integration.

But one aspect was ignored: performance of the generated code. First, component-based software was designed for distributed systems with interconnections of about 1MByte/s. Thus, the costs of argument marshaling were believed to be negligible because of the high costs of network communication. But this is not true for current network technology and definitely not true for current microkernel IPC. Thus, a large portion of the total cost of a function call between two threads is argument marshaling.

This paper describes ideas for an IDL compiler using detailed knowledge of the underly-ing communication mechanism to optimize argument marshaling. Furthermore, we allow the user to influence the code-generation process with additional meta information about involved components. This work is motivated by experiences we gained using an existing IDL compiler (*Flick* [3]) to build a Multiserver File System on top of the L4 microkernel [12].

2 Towards fast IDL systems

One of the first projects which dealt with this problem of building a fast IDL system was *Flick* by the University of Utah [3]. Its aim is to build a highly flexible IDL compiler which can be used with various IDL types as well as generate code for different communication platforms. One of its major ideas to improve the performance of argument marshaling is to enable the native language compiler (i.e., the C compiler) to optimize the marshaling code by function inlining instead of using separate marshaling calls. The work showed that runtime overhead of marshaling and unmarshaling can be reduced significantly by using inlined code.

But Flick still shares some drawbacks with other IDL systems. One of those drawbacks is the limited ability to optimize marshaling towards specific communication platforms. Communication is based on a separate message buffer for the arguments. Fig. 2 shows how this traditional argument marshaling works for a function call between two components in separate address spaces. The arguments of the function call are copied into a communication buffer in the sender components address space, this buffer is copied by the kernel communication primitive to a communication buffer in the receiver component and then to the argument buffers of the server side function.

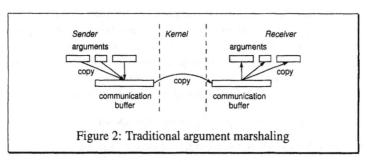

Figure 2: Traditional argument marshaling

This method involves three copy operations, although only one is necessary. The copy between the two address spaces by the kernel is necessary to uphold memory protection. The separate communication buffers are not required in case of a communication between components on the same host. Instead, the arguments can be copied directly from the client buffers to the server buffers, causing only one copy operation.

But copy operations are still expensive, especially for large amounts of data. Instead of copying the data, the client buffer can be shared between client and server.[1] That eliminates the copy operation, but causes additional costs for establishing the mapping. Those costs depend very much on the microkernel and hardware architecture. This requires the possibility to influence the generation of the marshaling code, e.g., to specify a threshold for the use of mappings instead of copying the arguments. But sometimes it is not even necessary to establish the mapping by the marshaling code, some systems already provide shared memory, which can be used to transfer the arguments. Again, this requires the possibility to specify the target environment to enable the IDL compiler to customize the code generation.

The ability to customize code generation is a very important requirement for the design of an IDL compiler for a microkernel-based system. The optimal marshaling method varies between different target architectures and system environments. This cannot be accomplished

---

[1] Assuming client and server trust each other

by a generic marshaling method. For example, modern microkernels use registers to transfer small amounts of data, but the exact number of available registers depends on the particular hardware architecture. A good example is scatter gather IPC, the ability to transfer data from/to scattered buffers (Fig. 3). The IDL compiler needs hints whether the target kernel is able to use this mechanism or not.

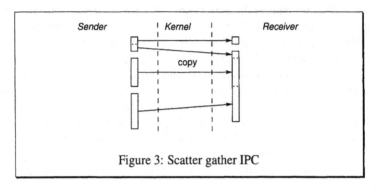

Figure 3: Scatter gather IPC

The optimal communication method also depends on the location of the communication partners. In case of intra-address space communication, arguments should be passed by references. This is not possible in the case of inter-address space communication. The IDL compiler should consider this by generating different implementations for the marshaling functions dependent on the location and trust of the communication partners.

To summarize, an IDL system for microkernel-based systems must target the following:

- Reduce the marshaling costs, especially avoid copy operations.

- Provide a flexible mechanism to customize the code generation.

- Generate different versions dependent on the location of the communication partners.

The next section describes the approach for a new IDL compiler which considers these requirements.

3   FIDL

Attacking the mentioned problems and incorporating our ideas we developed a new IDL compiler, *FIDL*. FIDL is based on an extended COM IDL. We chose COM for the following reasons:

- COM allows the specification of additional attributes of function arguments. Those attributes can be used to give hints to the IDL compiler, e.g., whether a string should be copied or transferred by reference.

- The COM type system is oriented to the C/C++ programming language. This provides more knowledge for the optimization of the marshaling code.

- COM does not dictate a copy in/copy out semantic.

Like Flick, FIDL creates the marshaling code as inline functions, thus enabling the C compiler to do the optimization. The IDL compiler itself can optimize the remote function call even further by exploiting the IPC mechanism of the underlying microkernel. As described above, it may be faster to establish a temporary mapping to transfer a larger amount of data rather than copying it to a separate communication buffer. But the exact threshold for that decision depends on the hardware architecture. Similar to COM-IDL, that kind of meta information can be provided in a separate *Application Configuration File (ACF)*. This file contains the following information:

- Hints for the IDL compiler, like the value of the threshold for using a temporary mapping.

- Information how the IDL compiler must marshal and unmarshal user defined data types. Those functions are used to transfer complex data type like linked lists, which cannot be handled by the compiler itself. This is similar to the *type translation information* of the Mach interface compiler [1] or the native data types in CORBA.

- Specialized implementations for the marshaling/unmarshaling or the IPC code. This can be used to optimize some or all function calls manually, e.g. to use existing shared memory areas.

- Functions to customize memory allocation and synchronization.

To summarize, the ACF contains all information to adapt the code generation to a particular system architecture.

The IDL compiler generates different versions of the client and server stubs, depending on the location of the communication partner. If the sender and destination threads reside in the same address space, arguments can be passed as memory references. If sender and receiver reside in different address spaces, the arguments must be copied or mapped. The proper implementation is assigned by a function table. This function table is created and initialized during the setup of a communication relation.

4 Measurements

To evaluate our ideas, we implemented a FIDL prototype. It generates the communication code for L4 IPC. Currently, it provides only a restricted functionality, it can only handle basic data types and arrays.

For our measurements we used a Pentium machine running the L4 microkernel resp. Linux (for rpcgen).

4.1 Marshaling costs

In our first test we measured the overhead caused by the argument marshaling. Fig. 4 shows the interface specifications for FIDL, Flick and *rpcgen*, the IDL compiler for SUN RPC [11].

All IDL compilers used a separate communication buffer. Table 1 shows the marshaling costs and additionally the costs for a hand-coded version of the argument marshaling.

The large difference between the overhead of rpcgen on the one and FIDL and Flick on the other side is mainly caused by two reasons:

```
SUN RPC:

struct msg {
  unsigned int a;
  unsigned int b;
  char c<20>;
};

program rpc_test {
  version testvers {
    void func(msg) = 1;
  } = 1;
} = 0x20000001;
```

Flick:

```
typedef string<20> string20;

module test {
  interface test {
    void func(in long a, in long b,
              in string20 c);
  }
}
```

FIDL:

```
library test {
  interface test {
    void func([in] int a, [in] int b,
              [in,size_is(20)] char *c);
  }
}
```

Figure 4: IDL sources

IDL compiler	Marshaling costs (cycles)
rpcgen	3275
Flick	471
FIDL	248
hand coded	161

Table 1: Marshaling costs

1. rpcgen creates a hardware independent representation of the arguments (XDR). This is required in distributed systems where components run on different hosts.

2. rpcgen uses separate marshaling functions rather than inlining the code to the function stubs.

4.2 End-to-End communication

The more interesting numbers are the costs of the actual function call. Those costs include the argument marshaling and the IPC call. Fig. 5 shows the times of a function invocation depending on the argument size for Flick and different versions of FIDL.

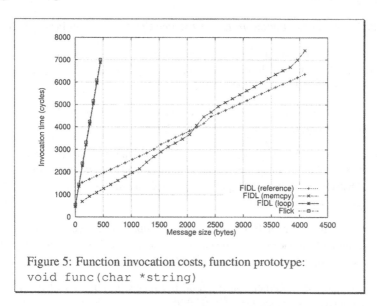

Figure 5: Function invocation costs, function prototype:
`void func(char *string)`

The three versions of FIDL use different methods to marshal the function arguments. *FIDL(loop)* uses the same implementation like Flick, thereby each character of the string is copied to the message buffer in a loop. *FIDL(memcpy)* uses the `memcpy` function to copy the string to the message buffer. This function is much better optimized than the loop in the first case. *FIDL (reference)* does not use a separate message buffer, instead it passes a reference to the string to the microkernel, the microkernel copies the string directly from the original buffer. As explained above, this eliminates one copy operation, but introduces a larger overhead in the IPC communication as the measurement results in Fig. 5 show.[2] This confirms our claim, that a very flexible IDL compiler is required, which for example generates different different marshaling code for strings depending on the size of the string.

5 Outlook

FIDL combines established and well understood technology, IDL compilers, with new ideas for the optimization of communication between components. Our initial performance is promising.

A fast communication mechanism is a key requirement for the success of a microkernel-based system. However, communication is only one basic mechanism in a component system. More *services* are required on top of those mechanisms such as the *CORBA Services* [8]. Further work is required to provide a complete infrastructure for supporting the development of systems such as *DROPS* [5] on top of microkernels.

---

[2]The overhead is caused mainly by a more complex setup of the copy operation in the kernel.

References

[1] Richard P. Draves, Michael B. Jones, and Mary R. Thompson. MIG — the MACH Interface Generator. Unpublished manuscript from the School of Computer Science, Carnegie Mellon University.

[2] Guy Eddon and Henry Eddon. *Inside Distributed COM*. Microsoft Press, 1998.

[3] Eric Eide, Kevin Frei, Bryan Ford, Jay Lepreau, and Gary Lindstrom. Flick: A Flexible, Optimizing IDL Compiler. In *Proceedings of the Conference on Programming Language Design and Implementation (PLDI)*, 1997.

[4] D. Golub, R. Dean, A. Forin, and R. Rashid. Unix as an Application Program. In *USENIX 1990 Summer Conference*, pages 87–95, June 1990.

[5] Hermann Härtig, Robert Baumgartl, Martin Borriss, Claude Hamann, Michael Hohmuth, Frank Mehnert, Lars Reuther, Sebastian Schönberg, and Jean Wolter. DROPS - OS Support for Distributed Multimedia Applications. In *Proceedings of the Eigth ACM SIGOPS European Workshop*, 1998.

[6] Hermann Härtig, Michael Hohmuth, Jochen Liedtke, Sebastian Schönberg, and Jean Wolter. The Performance of μ-Kernel-Based Systems. In *Proceedings of the 16th ACM Symposium on Operating System Principles (SOSP)*, 1997.

[7] Jochen Liedtke. On μ-Kernel Construction. In *Proceedings of the 15th ACM Symposium on Operating System Principles (SOSP)*, 1995.

[8] The Object Management Group (OMG). *The Complete CORBAServices book*. http://www.omg.org/library/csindx.html.

[9] Alan Pope. *The Corba Reference Guide: Understanding the Common Object Request Broker Architecture*. Addison-Wesley, 1998.

[10] M. Rozier, A. Abrossimov, F. Armand, I. Boule, M. Gien, M. Guillemont, F. Herrmann, C. Kaiser, S. Langlois, P. Leonard, and W. Neuhauser. CHORUS Distributed Operating System. *Computing Systems*, 1(4):305–370, 1988.

[11] R. Srinivasan. RPC: Remote Procedure Call Protocol Specification Version 2. Technical report, Sun Microsystems Inc., 1995.

[12] Volkmar Uhlig. A Micro-Kernel-Based Multiserver File System and Development Environment. Technical Report RC21582, IBM T.J. Watson Research Center, 1999.

HIGH-LEVEL SYSTEM SYNTHESIS

Application Cases

Hardware/Software-Architecture and High Level Design Approach for Protocol Processing Acceleration

Mirko Benz, Georg H. Overbeck
Department of Computer Science
Dresden University of Technology
D-01062 Dresden, Germany
{benz, overbeck}@ibdr.inf.tu-dresden.de

Klaus Feske, Jens Grusa
FhG IIS Erlangen
Department EAS Dresden, Zeunerstr. 38
D-01069 Dresden, Germany
{feske, grusa}@eas.iis.fhg.de

Abstract

Developing hardware support for transport layer protocol processing is a very complex and demanding task. However, for optimal performance hardware acceleration can be required. To cope with this situation we present a high level design approach which targets the development of configurable and reusable components. Therefore we outline the integration of advanced tools for the development of controller systems into our design environment. This process is illustrated based on a TCP/IP header analysis and validation component for which initial performance results are presented. The development of these specialised components is embedded in an approach to develop flexible and configurable protocol engines that can be optimised for specific applications.

1 Introduction and Related Work

Today's communication environments are mainly influenced by the tremendous success of the Internet. As a result the Internet Protocol (IP) and standard layers above - especially TCP [18, 19] - are now the common denominator. This means that although these protocols have a number of limitations concerning functionality, flexibility and performance other protocol approaches like XTP [17] have failed to gain broad acceptance. This is also partly true for other superior technologies like ATM which compete with IP. Hence it is important to transfer the alternatives and ideas developed in various research projects to improve implementations of these standard protocols.

On the other hand, the Internet has encouraged huge investments in fibre optical networks and technologies to exploit them more efficiently like Wave Division Multiplex (WDM). Furthermore, new technologies like xDSL and cable modems will also provide high speed communication in the access networks. Altogether this will contribute to an emerging global high speed networking infrastructure based on IP.

In contrast to using the same base protocol everywhere, communication devices are extremely diversified. This includes standard workstation and server class computers as well as laptops up to Wireless Application Protocol (WAP) mobile phones. Therefore, architectures for protocol processing acceleration have to be adaptive to various network interfaces, their properties as well as processor architectures or optimisation goals concerning performance as well as memory, CPU and power limitations. Furthermore, the ongoing development within and the commercialisation of the Internet have produced quite a number of applications, protocols and service proposals and standards for higher layers or extensions to IP. Examples are IP security for virtual private networks (VPN), voice over IP, video conferencing or the WWW. Especially the real-time requirements of multimedia data transmission stimulated the

R. Merker and W. Schwarz (eds.), System Design Automation, 75-86.

development of resource reservation protocols, priority mechanisms, accounting or the differentiated services approach.

The ongoing research and deployment of WDM technologies and the direct transmission of IP datagrams over specific wave lengths, will contribute to very high bandwidth capacities at low error rates. These optical networks will again shift the protocol processing overhead to the access routers and into the end systems. On the other hand they will probably provide no or only limited quality of service (QoS) features. Combined with data touching intensive or real time requirements of specific services this adds to processing power that is required within endsystems. Hence, flexible architectures for protocol processing acceleration are necessary to cope with these conditions.

There have been quite a number of approaches for hardware support of communication protocols in the early nineties. However they were only successful for lower layer protocols (e.g. MAC sublayer) [1, 13, 9]. The most significant aspect is probably the complexity of standard communication stacks like TCP/IP. This makes it impossible to perform all the processing in custom hardware because the design, implementation, testing, validation and maintenance effort would be too high. This would also lead to extreme costs and limit the ability to adopt standard modifications and improvements. Another obstacle is the lack of a formal specification. Hence, many research projects had concentrated on the hardware support of specialised light weight protocols [8, 15]. Although this was successfully demonstrated, hardware support for complex transport level protocols is still an open issue.

On the other hand, hardware support for communication protocols is again a very active topic. Especially so-called network processors are very popular. They include custom hardware for standard protocol specific computations as well as multiple programmable RISC cores, which makes them more flexible than custom ASIC designs that are used in routers today. This approach provides benefits concerning time to market and allows to continue development after purchase and to update the devices as required. This development is supported by improved tool support for simulation, verification and synthesis, because of the expansion of the component idea onto the hardware design in the form of intellectual property cores and subsequent reuse as well as higher abstraction levels like hardware compilation approaches. An example is the integration of high level design tools [20] in the process of developing hardware components. This is shown in [16, 7] and can improve productivity. Furthermore ASIC and especially FPGA technologies for hardware prototyping were drastically improved. Combined, these achievements facilitate the design and implementation of complex and heterogeneous hardware/software architectures for protocol processing acceleration and system-on-a-chip solutions. One such example is the hardware support for ATM, which often integrates transport layer functionality like segmentation and reassembling, congestion control (ABR) or traffic shaping in hardware.

In the following chapter we outline our protocol engine approach. First we state design goals and present our general approach. Then we outline the aspired architecture and discuss possible configurations and describe the involved components. In chapter 3 we illustrate a possible TCP/IP partitioning, explain the required protocol processing and outline a synchronisation between the software and hardware parts. Then, we describe our validation architecture for a hardware implementation of TCP/IP core functionality and present initial evaluations and performance data.

2 Protocol Engine Project

The protocol engine project is a joint effort of multiple research groups with a computer science and electrical engineering background. In this context communication protocols are analysed, evaluated and optimised. Specific protocols tailored for ATM networks and multimedia applications have been developed. For existing standard protocols hardware support is evaluated and designed. The focus of this paper is to describe the basic approach and architecture of the protocol engine with hardware support for TCP/IP in mind as well as a prototype implementation. The overall goal however, is to see networking in its entirety - ranging from applications, over protocols to the actual hardware. This way, by not looking at one specific layer alone, system performance can be improved.

2.1 Assumptions, Approach and Design Goals

Today's transport protocols like TCP are far too complex to be completely implemented in custom hardware. Furthermore this approach would limit the flexibility and maintainability of the solution. On the other hand, only a very small part of the protocol has real-time processing requirements. The rest consists of lower priority tasks like exception handling, buffer registration or connection management. This means, that only a relatively small part of the protocol has to be accelerated.

Within modern local high speed networks the error probability is very low. Hence exception handling due to corrupted data, loss or duplicates can be considered as a rare condition. As a consequence, non real-time processing like connection setup or exception handling in case of errors can still be performed by a modified software stack because relatively expensive synchronisation is tolerable. This way, investments in high performance TCP/IP implementations can be reused. Another advantage is that a step wise optimisation - based on existing implementations - and tailored to specific requirements is possible.

Advances in CPU development have greatly improved the processing power that is available in endsystems. However, due to added services which demand very high processing capacity or exhibit very tight real-time requirements hardware support could still make sense. Another issue is that the operating system becomes a bottleneck for simple protocol processing tasks because of context changes, synchronisation or the communication overhead inherent in layered architectures relative to the protocol processing itself. Hence the data path between the application and the network is very important too. From these requirements and constraints we derive the following design goals:

- Flexible, adaptive architecture to support different protocols based on IP. Allow easy extensions to support new features and services.
- Scalable performance according to specific requirements like cost, power consumption and network conditions.
- Development of components for specific protocol functions. Design of common interfaces to allow reuse, hardware emulation in early design phases and flexibility concerning the implementation architecture.
- Allow integration of specific hardware to benefit from existing solutions or for very high performance requirements. Furthermore enable easy integration of additional processing resources like DSPs or micro controllers.
- Take advantage of existing software based protocol implementations and allow stepwise hardware support.

- Consider protocol processing as a whole. This means that the network interface, the protocol processing and the communication with the application have to be regarded and optimised in its entirety.
- On wire compatibility to other (software based) implementations.
- Enable existing applications to take advantage of the protocol processing acceleration in a transparent manner.

2.2 Architecture and Configuration Opportunities

Due to the number of requirements and environmental conditions a protocol engine has to be designed specifically to address these challenges. As a consequence there will be many configurations. The general idea however is to add hardware support as required, to allow a smooth upgrade path and enable scalability. Figure 1 presents the overall architecture of the protocol engine with all possible components.

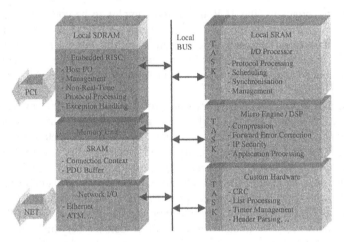

Fig. 1: General Protocol Engine Architecture

Depending on the specific application requirements there can be various configurations of this protocol engine. One opportunity consists in using currently emerging network processor designs which contain a standard embedded RISC processor combined with multiple programmable micro engines and very few network specific hardware like the LEVEL ONE IXP1200 architecture [11]. These architectures are especially suited for technology demonstrations of emerging standards like the integrated services approach. Because they are fully programmable, specification modifications can be quickly adopted. Furthermore they offer relatively high performance.

Another direction would be to leave the majority of the protocol processing tasks on the host and support only the standard data path with specific hardware. The high bandwidth of today's networks combined with extremely low error rates makes this approach feasible. As a consequence only in the rare case of exceptions due to error or connection management synchronisation handling would be required. Supported by a high level design tool a scheme for the development of such protocol accelerators was outlined in [4] for TCP/IP packet validation within the receive path.

For greater flexibility we plan the integration of a specific I/O processor [5]. This processor will possess a customisable instruction set. Hence it can be optimised for stream parsing and bit operations for example. This can lead to a reduction of code size and required processing cycles which is a general goal for power consumption sensitive applications.

On the other hand we plan to use the I/O processor to control the operation of the protocol engine. This means that it can be programmed to control the processing flow. As such it must communicate with the other entities of the protocol engine. Hence it must have knowledge of the number and capabilities of the integrated components and optimise their utilisation. We plan to develop a message abstraction layer for the exchange of such processing requests (TASK). This enables the I/O processor to communicate with other processing elements without having to know their implementation architecture in an asynchronous fashion. Therefore, this adaptation contains a common message buffer and a bus access part as well as a specific interface to the component. This way, specific hardware could be emulated during early stages of the development process. This could be a very high performance DSP or protocol specific hardware. On the other hand this abstraction opens the way to scalability since multiple components of the same type could be integrated as well. In this configuration the I/O processor would be responsible to synchronise the processing results of the entities and communicate with a standard embedded RISC processor.

Due to the complexity of transport layer protocols it is usually not beneficial to perform the entire protocol processing in specialised components. Hence we intend to use a general processor which runs a modified software TCP/IP stack. Here non real-time processing tasks like connection management or exception handling are performed. Furthermore, it is responsible for efficient communication with the application on the host system. To avoid operating system overhead we consider to use a modified implementation of the virtual interface architecture (VIA) for efficient communication [21]. Further information, for example how applications could transparently – without modifications – benefit from this acceleration can be found in [2, 3].

3   TCP/IP Partitioning and Fast Path Processing

Judged by the lines of code TCP is a relatively complex protocol [12]. However, assuming bulk data transfer within local area networks only a fraction of the code is actually required to process most packets. This is called the fast path. This state is reached after the connection is established and holds on as long as normal data packets with no control flags in the header are transmitted. A further requirement is the absence of error conditions due to loss, congestion or data corruption. Most of these conditions are relatively rare within today's local high speed networks. To achieve good performance it is therefore necessary to optimise the fast path. As a consequence it is not necessary or beneficial to implement complex protocols completely in hardware but to only support the common path with specific accelerators. Hence the majority of the protocol will still reside in software.

Within the fast path the sending instance checks whether available data can be transmitted. This is for example the case if the transfer unit of the network can be fully utilised. Large messages can be fragmented or small ones are aggregated. Thus, message boundaries are not preserved. The header fields are set and a checksum is computed then IP is invoked for the actual transmission. At the receiver, fast path processing consists of header analysis, context lookup, checksum calculation, packet validation, packet reassembling and hand over to the higher layers. Important optimisations include header prediction, context caching and integration of checksum computation in buffer copy operations [10, 14]. As a consequence of these optimisations it normally takes very few instructions to process a packet. Assuming faster and faster processors while the protocol processing remains essentially the same the ratio between actually required processing and overheads is getting worse. Since protocol processing is not the bottleneck (if not additional services like IP security are used) user mode implementations present an alternative. However, without intelligent hardware support they can not fully exploit today's networks.

3.1 Fast Path Protocol Processing

Figure 2 illustrates the tasks that are involved in the fast path processing for bulk data transmission. The sender accepts the application's data and transmits them if certain criteria are met. The receiving protocol instance takes and validates the data and eventually hands them over to the application. Both protocol instances are coupled by a window based flow control that ensures that enough buffer space is available at the receiver. Usually for every other received protocol data unit the receiver generates an acknowledgement. Based on this information the sender can release transmitted data that was saved for eventual retransmission. Furthermore, the transmission window is enlarged enabling the transmission of new data.

The fast path consists of three major components: TcpSend, TcpRecv and SendAck. These tasks may run concurrently but since they access shared context data, synchronisation has to be applied. Furthermore, they signal each other required processing like the receiver indicating necessity of sending an acknowledgement. Depending on the communication behaviour only some tasks are active on each instance. The context data include information describing the connection and variables for flow and congestion control as well as for error detection. These data sets are kept separate for every connection. Hence connections can be processed concurrently. Statistics data however, are gathered for every connection and thus have to be periodically synchronised with the software stack.

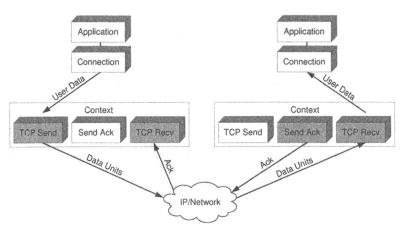

Fig. 2: TCP Fast Path Processing Flow for Bulk Data Transfer

3.2 Software Stack Synchronisation

According to the proposed partitioning the complex tasks of connection management and error handling are still performed in software. Thus, we can benefit from existing well performing and stable implementations. Therefore, when an application initiates a connection establishment the software stack is invoked. On the receiving side the protocol data unit can not be mapped onto a known fast path entry, therefore it is passed to the software stack. If the user decides to accept the communication the context information for this connection is transferred to the fast path processing unit and marked active. The same happens on the client side. The number of connections that should be accelerated may be limited to participants of the same or corresponding networks because normally only here very high performance is required and the above mentioned conditions are met. The data units that are exchanged afterwards are entirely processed by this unit and transferred directly to the application. This happens without invoking the operating system or the TCP/IP software stack. This means that no interrupts have to be dealt with, no operating system contexts or processor modes have to be changed and caches could remain intact.

When a connection is idle for some time, errors occur or the user terminates it, the context information and remaining user data are transferred to the software stack. Then the communication is treated as it would have been without an accelerator. In case an error condition was successfully managed, the fast path unit could be reinitialised for this connection.

4   A TCP/IP Receive Side Processing Component

Hardware support and parallelism to improve protocol processing performance were not considered within the specification of TCP/IP. As a consequence, there are a lot of dependencies between protocol functions, shared access to the connection state data (the transmission control block) and a high communication and synchronisation effort between protocol functions. To make matters worse, this mainly depends on the actual transported data. Therefore, a functional decomposition of the protocol is a difficult task. Hence, a prior analysis and simulation of the entire system is necessary to achieve good results.

In this section we describe how hardware support can be integrated within the receive path processing of TCP/IP. Again, we assume low error rates and high network bandwidth. Therefore, the required processing is mainly to assure error free data reception and hand over to the user process. However, this has to be done very efficiently to cope with gigabit data rates.

4.1 Simulation Model and Test Environment

Figure 3 represents the simulation test bench currently in use. The major goals were prototyping of the fast path unit, exploration of different configurations, bottleneck investigation and derivation of initial performance data. For this to be accomplished, the interface to the network (e.g. MAC in the case of Ethernet) is currently emulated. Furthermore, only one connection is actively processed.

All models except the interfaces to the application and to the remote TCP/IP instance are written in VHDL. The SRam model allows to dynamically update its contents via a file

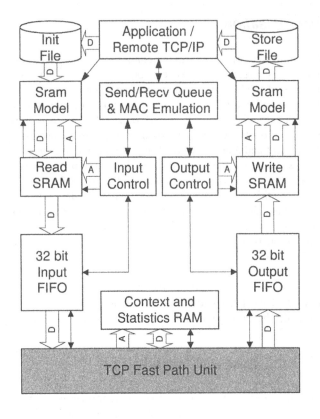

Fig. 3: Simulation Model

interface. This is for example used by the application interface when a packet should be transmitted. First the user data is transferred to the init file. Then the contents of the SRam is updated. The position and length of this data is stored in a transmission queue that is available for each connection. Via this queue a synchronisation between the software application and the hardware unit is performed. The input control generates a header and transfers the described buffer to the Input FIFO. The header contains a packet type field. This could be application (send/recv), network (send/recv) as well as synchronisation and statistics. Furthermore, it contains the length and packet specific data - for example the connection to which this packet belongs or the network interface that received the packet. After the packet is transferred the queue entry is cleared and can be reused by the application. The fast path unit is invoked if a configurable threshold for the Input FIFO filling is reached. The processing of this unit is described in the next section. Here a local SRam can be utilised to store connection and statistic information. The data extraction from the Output FIFO and the transfer to the application or network emulation works in a similar manner. The models for FIFO elements and the local Ram target Block Select Ram of Xilinx Virtex FPGA devices.

4.2 TCP Fast Path Unit Receive Processing Flow

Figure 4 represents the general processing flow within the TCP fast path unit. After leaving the reset state the unit communicates with the input FIFO and waits for available data. It then checks whether the received packet contains synchronisation information. If this is the case, the connection context data is initialised. Since currently only the receive path is implemented a filter for those packets is inserted within the processing flow. Other packets are directly transferred to the output FIFO. The next step consists of extracting the connection description (IP addresses and TCP ports) and a comparison with the currently loaded context information. If the packet does not belong to the accelerated connection it is simply forwarded. Otherwise it is analysed as explained in section 3.1. Within the data validation step the TCP checksum is computed for the entire packet. If this succeeds the data is transferred to the corresponding application. Furthermore an acknowledgement packet for the received data is generated.

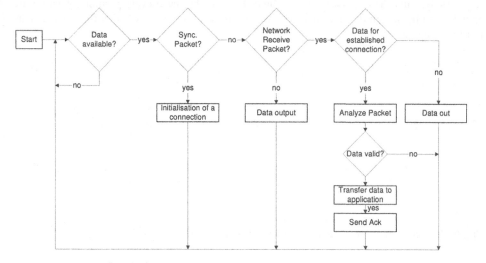

Fig. 4: TCP Fast Path Unit Processing Flow

4.3 Design Flow, Implementation and First Results

Conventionally, the controller design of structured data stream processing is not well supported by EDA tools. So we are facing a bottleneck in the design process especially for protocol processing hardware components. To fill the gap, we incorporated modelling and synthesis facilities of the Protocol Compiler from Synopsys [20] into a proved FPGA based rapid prototyping (RPT) design flow [6] and utilised it for designing fast path protocol processing components. Our RPT design flow starts at RT-Level with a VHDL description of the design specification. As a new element in the sequence of this flow, the Protocol

?	unspecified frame
G	action (default, user)
⊡	terminal frame
R	reference frame
Z	epsilon operator
⊡	qualifier operator
⊴	if-frame operator
⊡	repeat operator
()	sequential operator
‖	alternative operator

Fig. 5: Types of frames and frame operators

Compiler is set on top of the whole process, by means of which the high level specification is graphically composed. Furthermore the Protocol Compiler provides the following features: formal protocol analysis, back annotation simulation, controller logic partitioning and synthesis, and VHDL code generation. Being similar to the Backus-Naur-notation the Frame Modelling Language FML [20] and the graphical symbolic format (Figure 5) closely match requirements of high-level protocol specifications like

- Recognition of header patterns and synchronisation,

- Parsing and reassembling of structured data streams,

- Interface issues between data stream processing modules (such as synchronisation or stall).

As an example Figure 6 shows the Protocol Compiler description of a simplified 32 bit IP header analysis implementation. A 32 bit wide register p_data_in is used to processes the data sequentially. First the checksum is computed. Here, within each cycle portions of 16 bit are added. Then the length of the IP header is extracted. Next we check whether the received data is a IPv4 fragment.

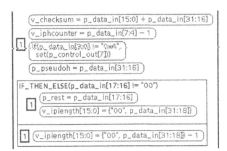

Ver	IHL	TOS	Total Length	
IP Identification			Flags	Frag. Offset
TTL		Protocol	Header Checksum	
Source IP Address				
Destination IP Address				

Fig. 6: Protocol Compiler IP Header Analysis Specification (excerpt, 32 bit)

Within the next step the pseudo header is initialised which is required for the checksum computations of higher layers. The next step calculates the number of 32 bit words (v_iplength) that have to be processed and determines eventually remaining bytes (p_rest). All these actions are performed within one clock cycle. Than the rest is read in units of 32 bytes. Within these steps header fields like the source IP address are extracted. Furthermore option fields are taken into account. After the header fields are extracted we perform validation operations as outlined in the previous section to verify the received fragment.

One part of the top-level is shown in figure 7. It illustrates three alternatives of processing after receiving valid data. The first four bits of the incoming packet decide which of the alternatives will run. If DIN is equal to „5" (decimal) the following data is used to initialise a new connection. The packet is transferred directly to the output-FIFO if DIN is „3", because only the receive path is implemented at this time (see part 4.2). A normal analysis of a packet begins only if DIN is equal to „1". After starting the packet analysis by checking the Ethernet MAC address referenced to „ethernet_input" the fast path unit parses the IP-Header and the TCP-Header. The last step is the generation of an acknowledgment if the encapsulated data in the packet was valid.

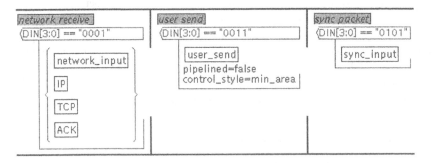

Fig. 7: Fast Path Unit Top Level Protocol Compiler Description

The Protocol Compiler offers the possibility to synthesize the design applying certain optimisation criteria. For example in figure 5 the attribute „Control_Style" is set to „Min Area". That means the partition „user_send" will be optimised for minimum area during high level synthesis. Table 1 shows results of the synthesis for a Xilinx Virtex FPGA XV300 device and the differences between the Control_Style „Min Area" and „Min Delay".

Table 1: Synthesis Results

Option Property	Minimum Area	Minimum Delay
Maximum Frequency	19,1 MHz	22,3 MHz
Slices	967	982
External I/O (IOBs)	71	71

5 Status and Future Work

We have presented a concept for the design and implementation of protocol processing accelerators. This was demonstrated with a packet classification and validation unit for hardware supported TCP/IP receive path processing. First steps in modelling and synthesis of the hardware partition where made utilising a usual FPGA rapid prototyping design flow, which we extended by a graphical high level design entry and by reusable protocol components. This approach supports an application-oriented modelling style in order to enhance design efficiency and quality for structured data stream processing controllers. Additionally, this leads to a quick design exploration, easy changeability, and design cycle reduction. A further controller design improvement can be expected by utilising reuse methodologies. Consequently, our future work aims at extending our rapid prototyping design flow by inserting a library of reusable protocol templates and components.

The design of accelerators based on these components is relatively difficult because a number of conditions have an influence on the achievable performance. Due to the inherent complexity of transport layer protocols it is furthermore advantageous to perform the non real-time processing on standard programmable architectures. Furthermore we will evaluate the integration of signal processors as well. As a consequence there are a lot of design alternatives combined with protocol engine configurations and requirements to take into account. Hence an automated and integrated design approach to determine which architecture is best suited for a specific protocol processing task would be beneficial. Further issues would be performance prediction as well as simulation and validation of the entire communication system.

6 References

[1]   Balraj, T.S.; Yemini, Y.: "Putting the Transport Layer on VLSI - the PROMPT Protocol Chip", in: Pehrson, B.; Gunningberg, P.; Pink, S. (ed.): Protocols for High-Speed Networks, III, North Holland, Stockholm, May 1992, pp. 19-34

[2]   Benz, M.: "The Protocol Engine Project - An Integrated Hardware/Software Architecture for Protocol Processing Acceleration", SDA '2000 workshop

[3]   Benz, M.; Engel, F.: "Hardware Supported Protocol Processing for Gigabit Networks", SDA - Workshop on System Design Automation, 1998

[4]   Benz, M.; Feske, K.: "A Packet Classification and Validation Unit for Hardware Supported TCP/IP Receive Path Processing", SDA '2000 workshop

[5]   Engel, F.; Nuehrenberg, J.; Fettweis, G.P.: "A Fast and Retargetable Simulator for
        Application Specific Processor-Architectures", SDA '2000 workshop

[6]   Feske,K.; Scholz,M.; Doering,G.; Nareike,D.: "Rapid FPGA-Prototyping of a DAB
        Test Data Generator using Protocol Compiler", FPL'99, August 30th-Sept 1st 1999,
        Glasgow

[7]   Feske, K.; Döring, G.; Scholz, M.: "Efficient Design of Structured Data Processing
        Controllers Using Protocol Compiler and Behavioural Reuse - a Case Study.",
        accepted for DATE'2000, Paris, France, 27-30 March 2000

[8]   Krishnakumar, A.S.: "A Synthesis System for Communication Protocols", Proceedings
        of the 5th Annual IEEE International ASIC Conference and Exhibition, Rochester,
        New York, 1992

[9]   Krishnakumar, A.S.; Kneuer, J.G.; Shaw, A.J.: "HIPOD: An Architecture for High-
        Speed Protocol Implementations", in: Danthine, A.; Spaniol, O. (ed.): High
        Performance Networking, IV, IFIP, North-Holland, 1993, pp. 383-396

[10]   Koufopavlou, O.G., Tantawy, A.N., Zitterbart, M.: "Analysis of TCP/IP for High
        Performance Parallel Implementations", 17th IEEE Conference on Local Computer
        Networks, Minneapolis, Minnesota, September 1992

[11]   Level One™ IXP1200 Network Processor, Product Brief, www.level1.com, 1999

[12]   Microsoft Research IPv6 Implementation, www.research.microsoft.com/msripv6,
        1999

[13]   Morales, F.A.; Abu-Amara, H.: "Design of a Header Processor for the Psi
        Implementation of the Logical Link Control Protocol in LANs", 3rd IEEE
        International Symposium on High Performance Distributed Computing, San Francisco,
        1994, pp. 270-277

[14]   Pink, Stephen: "TCP/IP on Gigabit Networks, High Performance Networks, Frontiers
        and Experience", Kluwer Academic Publishers, 1994, pp 135 -156

[15]   Schiller, J.H.; Carle, G.J.: "Semi-automated Design of High-Performance
        Communication Systems", Proceedings of the 31st Annual IEEE International
        Conference on System Sciences, HICCS, Hawaii, 1998

[16]   Seawright, A. et al.: "A System for Compiling and Debugging Structured Data
        Processing Controllers", EURO-DAC'96, Geneva, Switzerland, Sept. 16-20, 1996

[17]   Strayer, W.T.; Dempsey, B.J.; Weaver, A.C.: "XTP - The Xpress Transfer Protocol",
        ADDISON-WESLEY, 1992

[18]   Stevens, W.R.: "TCP/IP Illustrated, Volume 1, The Protocols", ADDISON-WESLEY,
        1994

[19]   Stevens, W.R.; Wright, G.R.: "TCP/IP Illustrated, Volume 2, The Implementation",
        ADDISON-WESLEY, 1995

[20]   SYNOPSYS: "V1998.08 Protocol Compiler User`s Guide", Synopsys Inc., 1998

[21]   Virtual Interface Architecture, specification version 1.0, www.viarch.org, 1999

A Fast and Retargetable Simulator for Application Specific Processor Architectures

Frank Engel, Johannes Nührenberg and Gerhard P. Fettweis
Mannesmann Mobilfunk Chair for Mobile Communication Systems
Dresden University of Technology, 01062 Dresden, Germany

Abstract

Retargetability allows an easy adoption of a simulator on different processor architectures without a time consuming redesign of all tools. This is evident for an efficient HW/SW codesign.

In this paper we describe a tool set for fast and easy simulation of processor architectures based on a retargetable simulator core. This approach helps to reduce the development time for designing and validating System-on-a-chip (SoC) applications based on a processor core. The use of ANSI C avoids an expensive development of a modeling language.

Our main focus in this paper is on conceptual decisions we made and on the structure of the tool set.

1 Introduction

The progress in chip technology allows designers to create more and more complex designs known as System-on-a-chip (SoC) [4]. This means, systems like printed circuit boards so far assembled of discrete units (e.g. RS232 interface, memory controller, etc.) are now integrated on one silicon die. Synthesizable models of this units are required. Often they are configurable to allow a customization onto the application needs. Especially in the case of a processor core, modifying the instruction set leads to lower power consumption and smaller die size. These are major aspects in system design. Thus, the popularity of application-specific processors has raised constantly in recent years [8], [7].

Ways of modification are the customization of instruction set and architecture of existing standard cores [9] or the design of a new processor based on a modular instruction set architecture and customizable data paths [5].

Currently a lot of powerful tools like HDL simulators, silicon compilers or physical layout generators are available. But all of these tools address the HW related part of the design flow. The research for an automatic development of SW tools like an assembler, a simulation model or even a simulation environment is still at the beginning. Hence, this becomes more and more the critical path of the framework.

In this paper we present a fast and retargetable simulator for processor architectures. Together with an also retargetable assembler one can easily customize an existing processor toward system needs or find help on the design of a new core.

In our example, the target processor for concept validation was the In/Out-Processor (IOP) of the M3-DSP chipset [15]. This processor supports the DSP core by managing data transfer, data pre- and post-processing and interrupt-handling.

87

R. Merker and W. Schwarz (eds.), System Design Automation, 87-96.
© 2001 *Kluwer Academic Publishers.*

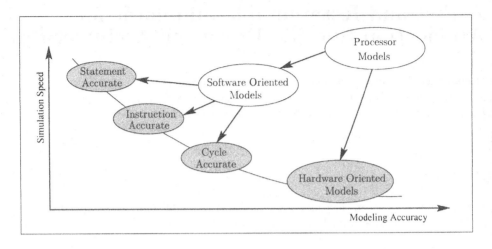

Figure 1: Types of Processor Models

The paper is structured as follows: In Section 2 we describe simulator concepts and explain basic decisions. The next section deals with a deeper description of main features. Section 4 resumes all parts to a design flow. Finally results are presented and an outlook on future work is given.

2   Simulator Concepts

A lot of Simulators or even emulators from different vendors are available for a specific processor. But they focus on fixed processors ready to use on a printed board. As depicted above, in the field of SoC customizable HW models are required. Thus, even the simulators need to be customizable. Unfortunately the number of research projects dealing with retargetable simulators is limited. They all use a special developed modeling language.

One example is MIMOLA developed at Dortmund University of Technology [1]. MIMOLA is a special language to describe the structure of a processor. It is similar to common hardware description languages (HDL) like VHDL or Verilog, but on a higher level of abstraction. The advantage is an easy generation of a synthesizable HDL description. Drawback is the low simulation speed similar to HDL simulators. To achieve fast simulations an extraction of a behavioral model is necessary. Unfortunately this extraction is restricted to a small class of fixed arithmetic DSPs.

Another example is Sim-nML of IIT, a language for behavioral description of a processor [3]. This is an extension of nML of Target Compiler Technologies [14]. The language allows the simulation of the instruction-pipeline. Although Sim-nML provides a compact description of a processor, the simulation speed is still limited [13].

The language LISA from Aachen University of Technology allows the modeling of the instruction set by dividing it into cycles [11]. The focus is on an exact pipeline representation. Thus, a cycle accurate simulation is implemented.

An easier and more efficient way we decided for is the use of an existing language like

ANSI C. There are a wide range of tools like compilers, linkers, debuggers for all platforms available. Furthermore it is a widely spread and well accepted programming language. We adapted it by defining a sufficient set of keywords for describing the instruction set and the processor behavior.

A simulator concept strongly depends on the required accuracy of the processor model. There are two major groups:

1. **HW oriented models** are used for designing structural models. The modeling of physical characteristics like gate-delays is required. Based on these descriptions synthesis tools generate gate-level structures for fabrication.

2. **SW oriented models** are needed for designing application software. Three subgroups can be defined:

   (a) *Statement level accuracy* is needed for testing influences on the algorithms. The focus is on problems like arithmetic overflow etc.

   (b) *Instruction level accuracy* is used for simulating the processor instructions to obtain benchmarks.

   (c) *Cycle level accuracy* is used for investigating correct timing. Often these models are used as a golden reference to validate the HW model.

As depicted in Figure 1 an increasing amount of modeling accuracy results in a decrease of simulation speed [17]. Hence, a tradeoff between maximum performance and the lowest acceptable accuracy level is required. A way to improve simulation speed is the use of compiled instead of interpreted simulation. Since the program for the target processor is a fixed sequence of instructions, some processor actions like the fetching and decoding of instructions can be eliminated. They are executed before the simulation run by translating them into a sequence of simulator statements. This is possible as long as the program running on the target processor is not self modifying. But since the majority of embedded applications run firmware stored in a ROM or an EPROM this type of programs can be neglected. An exception are data dependent instructions like conditional jumps. No preprocessing is possible. During simulation a jump to the succeeding instruction sequence is necessary.

For proof of concept our first target architecture was the IOP. As part of the M3-DSP chip set this processor is responsible for pre- and post-processing of data. The DSP-core itself is a highly parallel SIMD architecture which needs aligned data sets for full use of its processing power [15]. Thus one task of the IOP is the mapping of data streams onto aligned data sets for effective signal processing. Additionally this processor is responsible for scheduling of this data-stream manipulation together with asynchronous events like interrupts in a multitasking manner. Hence to simulate correct response times a cycle accurate model with full control of the pipeline behavior is required. Thus our simulator classifies to the cycle-level-accurate group in Figure 1.

To ensure platform independence we decided for Java and ANSI C as the programming and simulation languages.

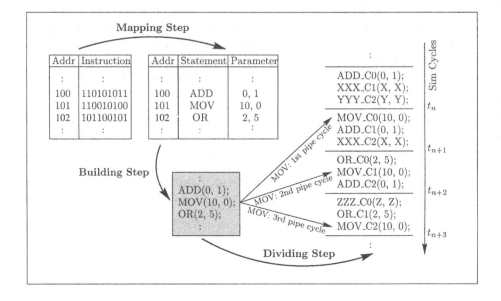

Figure 2: Steps of Compiled Simulation

3   Details

3.1   Compiled Simulation

The preprocessing of a compiled simulation is divided into the following steps (Figure 2):

1. **Mapping of instructions** and its parameters onto the corresponding simulator statements. This is equivalent to the decoding step within a processor. But instead of repetitive decoding each time the instruction is fetched, it is only done once. This is comparable to a reorganization of the instruction memory.

2. **Building a sequence of simulator statements.** This is possible since the common way of executing a program is the incrementing of the program counter. Thus the execution order of the instructions is known, except for conditional jumps.

3. **Dividing into pipeline steps and merging** with steps of other statements follows. This ensures the capability of simulating pipeline behavior.

3.2   Mapping of Concurrent Functionality

Besides the instruction decoder and the data path a processor contains separate units like timer, interrupt logic etc. All these modules are running in parallel from the instruction execution. Hence a mechanism to integrate them into simulation is needed. We introduced pre-cycle- and post-cycle-statements (Figure 3). These statements serve as an interface. Parameters are the processor status and the resources (registers, flags, etc.). Furthermore it provides reference points for co-simulation with other tools (Section 6).

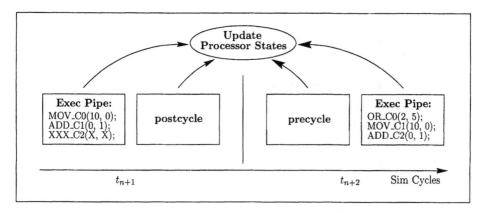

Figure 3: Functionality Between Cycles

A processor cycle is simulated as follows:

1. **Pre-cycle functionality:** The first interface is called. Due to the HW-supported multitasking in the IOP, this part is used to implement scheduling functionality.

2. **Execution of the pipeline:** During this stage pipeline steps of the currently running instructions are simulated. This represents the main activities in a simulation.

3. **Post-cycle functionality:** The second interface call is used to implement all autonomous modules (timer, etc.).

4. **Update processor states:** Each calculated result (like a register content) is temporary recorded in a change table. Now the writing of the new values is done. Since C is a sequential language this step emulates the concurrency of processes.

The change table is also used to observe multiple manipulations of variables (e.g. the content of a register). A warning message is generated if more than one entry for updating the same variable is found. This feature is helpful during the development of a new processor architecture.

3.3 Processor Description

To achieve retargetability the simulator is divided into two parts:

1. **Architecture independent:** This represents the frame of the simulator. It includes all routines for compiled simulation, handling of concurrent functionality, interfaces to debugger, etc. It is called the simulator base.

2. **Architecture dependent:** This part is derived from the Processor Description File (PD-File). A Java parser generates all necessary C routines from this information.

The PD-File is divided into the following main categories:

1. **Init:** This section defines global values necessary for simulation.

```
<init>
  int debug = 0;
<helpers>
  #include "helpers.h"
<intercycle>
<precycle>
  wait_for_user();

<instruction_set>

  <instruction>
    ADD

  <syntax>
    ADD Ry.Oy Rx.Ox

  <coding>
    [0]  [0]  [110000] 2222 1111 4444 3333

  <behavior>

    <cycle 0>
      SET_PC_ADDR_1;
      INC_PC(pc_addr);

    <cycle 1>
      SET_PC_ADDR_2;
      SHIFTER_Y(b0, b1, parameter[2]);
```

Figure 4: Example of a PD-File

2. **Helpers:** Here general C code can be directly included. This is useful for defining multiple used C routines and macros to describe the behavior of instructions.

3. **Resources:** This part defines all processor resources. The use of 'unsigned' C types is recommended to ensure a correct representation of bit vectors.

4. **Intercycle:** The concurrent functionalities are introduced here. Since they are directly linked into the simulator base they need to be implemented in ANSI C.

5. **Instruction_set:** The definition of instructions follows. This section is divided into the following subsections:

   (a) **Instruction:** This represents the mnemonic name of the instruction.

   (b) **Syntax:** The next line describes a disassembly string used in the debugger.

   (c) **Coding:** This section defines the semantic of each bit in the byte-code. Values in square brackets are fixed. Bits represented by the same numeral are grouped into one parameter. Furthermore this number denotes the parameter order in the assembler mnemonic. E.g. 2222 in Figure 4 denotes the second parameter Oy is a 4 bit value. Up to 9 parameters can be defined.

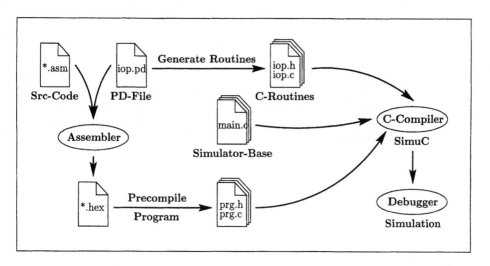

Figure 5: Simulation Flow

(d) **Behavior:** This part describes the behavior of an instruction. Here ANSI C code is required. Pipeline steps are represented by the key-word `<cycle n>`.

Figure 4 gives an example on how to describe the behavior of an instruction.

3.4 Assembler

Since the simulator requires a program in byte-code format, at least an assembler is required. This tool also needs to be retargetable. It uses the sections `<instruction>` and `<coding>` to generate a mapping scheme. All typical features of an assembler like, jump calculation, label resolving, generation of a listing, etc. are included. The tool comes with a graphical frontend for easy programming.

4 Simulation Flow

This section describes the complete tool set and the design flow as depicted in Figure 5. Assuming a complete processor description as a PD-File, the flow starts with writing assembler code.

In a second step the architecture dependent part of the simulator is generated from the PD-File. After that, according to Section 3.1, the preprocessing step of the compiled simulation is executed. It uses the byte-code produced by the assembler.

Now a standard C compiler is used to compile and link all three parts (simulator base, precompiled program and architecture dependent C routines) to an executable representing the simulator 'SimuC'.

Currently the free debugger 'DDD' is used as a graphical frontend during simulation [16].

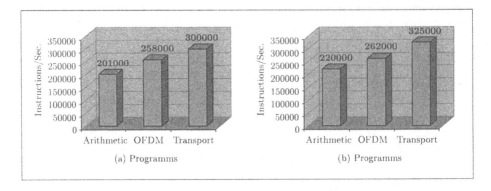

Figure 6: Results on Sparc (a), Intel (b)

5 Results

The tool set was developed on SUN workstations using the GNU compiler 'gcc', Java and the Java parser 'javacc'. Simulation speed around to 300,000 instructions per second are achieved. The performance is measured using different types of programs.

As depicted in Figure 6 the highest speed is achieved with programs mostly containing simple transport operations, up to 300,000 instructions per second on a SUN Sparc workstation (ULTRA1). On an Intel PC (PII-300) slightly more are achieved.

Since the IOP can operate on variable bits within a data element the cost intensive C modeling of bit operations (expensive bit masking on host architecture) results in a higher simulation effort. Thus a program containing arithmetic and logical operations on variable bit lengths needs longer simulation times.

A better benchmark is the OFDM-based Hiperlan wireless ATM modem, the target application of the M3-DSP chipset [6]. This program represents a profile of arithmetic/logical, transport and control instructions the IOP is designed for. This is comparable to results of other retargetable simulators (240,000 .. 300,000 [10], 160,000 [12]).

6 Summary and Future Work

In this paper we presented an easy approach for a retargetable processor simulator together with an assembler. Instead of designing a new modeling language to describe the processor architecture we choose ANSI C and Java. The advantage is a great variety of reliable tools like a compiler and the wide spread acceptance of these languages. The processor modeling is controlled by a set of keywords and constructs linked with self written C code (see Section 3.3). The user can still fully explore the capacity of ANSI C. Java is used for parsing the target program and the processor description.

Further investigations will deal with the design of a graphical frontend to replace the currently used DDD frontend. Also required is an interface for co-simulation with other tools like VHDL simulators or evaluation boards to integrate it into a complete system modeling environment [2].

References

[1] S. Bashford, U. Bieker, B. Harking, R. Leupers, P. Marwedela, A. Neumann, and D. Voggenauer. The MIMOLA Language Version 4.1. Technical report, University of Dortmund, 1994.

[2] M. Benz. The Protocol Engine Project. In *Workshop on System Design Automation (SDA)*, 13.-14. Mar. 2000.

[3] Cadence Research Center of Computer Science at the Indian Institute of Technology. Hompage. http://www/cse.iitk.ac.in/vrajesh/simnml.

[4] D&T Roundtable. Hardware-Software Codesign. *IEEE Design & Test of Computers*, 17(1):92–99, jan - mar 2000.

[5] G. Fettweis. DSP Cores for Mobile Communications: Where are we going ? In *IEEE International Conference on Acoustics, Speech and Signal Processing (ICASSP)*, volume I, pages 279–282, 1997.

[6] G. P. Fettweis, M. Weiss, W. Drescher, U. Walther, F. Engel, and S. Kobayashi. Breaking New Grounds Over 3000 MOPS: A Broadband Mobile Multimedia Modem DSP. In *International Conference on Signal Processing, Applications & Technology (ICSPAT)*, pages 1547–1551, 1998.

[7] L. H. Goldberg. Vendors Are Counting on Appliance-on-Chip Technology. *IEEE Computer*, 32(11):13–16, Nov. 1999.

[8] J. Hennessy. The Future of System Research. *IEEE Computer*, 32(8):27–33, Aug. 1999.

[9] K. Küçükçakar. An ASIP Design Methodology for Embedded Systems. In *Workshop on Hardware/Software Codesign (CODES)*, 1999.

[10] R. Leupers, J. Elste, and B. Landwehr. Generation of Interpretive and Compiled Instructions Set Simulators. In *Asia and South Pacific Design Automation Conference (ASP-DAC)*, 1999.

[11] S. Pees, A. Hoffmann, V. Zivojnovic, and H. Meyr. LISA - Machine Description Language for Cycle-Accurate Models of Programmable DSP Architectures. In *Design Automation Conference (DAC)*, 21.-25. June 1999.

[12] S. Pees, V. Zivojnovic, A. Ropers, and H. Meyr. Fast Simulation of the TI TMS320C20X DSP. In *International Conference on Signal Processing, Applications & Technology (ICSPAT)*, 1997.

[13] V. Rajesh. A Generic Aproach to Performance Modeling and Its Application to Simulator Generator. Master's thesis, Indian Institute of Technologies, 1998.

[14] Target Compiler Technologies. The Chess/Checker Retargetable DSP Environment. Hompage. http://www/retarget.com.

[15] M. Weiß, F. Engel, and G. P. Fettweis. A New Scalable DSP Architecture for System on Chip (soc) Domains. In *IEEE International Conference on Acoustics, Speech and Signal Processing (ICASSP)*, 1999.

[16] A. Zeller and D. Lütkehaus. DDD-A Free Graphical Front-End for UNIX Debuggers. Technical report, Braunschweig University of Technology, 1995. Informatik-Bericht No. 95-07.

[17] V. Zivojnovic. *DSP Processor/Compiler Co-Design: A Quantitative Approach.* PhD thesis, Aachen University of Technology, 1998. Shaker Verlag Aachen.

Hardware Supported Sorting: Design and Tradeoff Analysis

M. Bednara, O. Beyer, J. Teich, R. Wanka*
Universität Paderborn
D-33095 Paderborn, Germany
{bednara,beyer,teich}@date.upb.de, wanka@upb.de

Abstract

Sorting long sequences of keys is a problem that occurs in many different applications. For embedded systems, a uniprocessor software solution is often not applicable due to the low performance, while realizing multiprocessor sorting methods on parallel computers is much too expensive with respect to power consumption, physical weight, and cost. We investigate *cost/performance tradeoffs for hybrid sorting algorithms* that use a mixture of sequential merge sort and systolic insertion sort techniques. We propose a scalable architecture for integer sorting that consists of a uniprocessor and an FPGA-based parallel systolic co-processor. Speedups obtained analytically and experimentally and depending on hardware (cost) constraints are determined as a function of time constants of the uniprocessor and the co-processor.

1 Introduction

In this paper, we investigate the idea of combining a sequential sorting algorithm and a systolic hardware sorter. In particular, we analyze latency curves of both sorting techniques and derive an optimal solution for partitioning a sorting problem into hardware and software.

1.1 Sorting algorithms

A large family of sequential sorting algorithms is based on the concept of *merging*, i.e., on first recursively sorting two short subsequences and eventually producing a sorted overall sequence from the now sorted subsequences (see [2, pp. 158ff]). The sequential algorithm we use in this paper is merging-based. In order to sort N keys, merging-based algorithms need in the worst-case $N \log N + O(N)$ comparisons[1] and $\Theta(N)$ additional memory.

In parallel computing, several concepts were investigated. There is the hardware paradigm of *comparator circuits*, where the complete input is considered to be on the micro chip simultaneously. Most famous in this paradigm are Batcher's sorting circuits [1] that both sort $N = 2^d$ keys in $\frac{1}{2}d(d+1)$ parallel compare-exchange steps. Though they are not asymptotically optimal, these are the circuits that are the fastest sorters in applications.

The drawback of sorting circuits is that the time necessary to load the keys to the circuit is not accounted for. Often it takes time proportional to N to load the keys and to output the sequence after sorting. That means sorting needs $O(N)$ time although a sophisticated parallel sorting algorithm might have been used.

An approach that takes this into account is the *systolic* paradigm where sorting algorithms have been designed such that loading the keys and sorting are performed simultaneously. Leiserson's systolic algorithm [4] (a description of this algorithm can also be found in the overview paper [3]

---

*Supported in part by DFG Sonderforschungsbereich 376 "Massive Parallelität."
[1]Throughout this paper, log denotes the base 2-logarithm.

R. Merker and W. Schwarz (eds.), System Design Automation, 97-107.
© 2001 *Kluwer Academic Publishers.*

by Kramer and van Leeuwen) is the first systolic sorting algorithm. It works on a linear array of cells and is a variant of the sequential insertion sort (see [2, p. 80ff]). Here, the feeding of keys into the array is done at one end of the linear array. At the same time, the keys that are already in the array are subject to an ongoing sorting procedure such that after the last key has been loaded into the array, the first element of the sorted sequence can be output, then the next element, etc. Parhami and Kwai present in [5] a complete, optimized design for the cells, including the introduction of control bits that organize key loading, sorting, and output phases.

1.2   New contribution

In this paper, we present the design and evaluation results of a first hybrid implementation of a sorting algorithm that uses merge sort for its sequential part and Parhami's and Kwai's sytolic insertion sort for the parallel phases. We derive formulas for speedup with and without resource constraints on the number of PEs of the co-processors. The results are validated using a real FPGA implementation.

1.3   Organization of paper

In Sec. 2, the two plain sorting algorithms are presented. The combination of these algorithms in an hybrid approach is described in Sec. 3, together with a careful runtime analysis. In Sec. 4, we propose a hardware realization of the hybrid algorithm, and in Sec. 5, we describe our prototype implementation and the obtained results.

2   The Plain Sorting Algorithms

2.1   Sequential Merge Sort

Concerning the sequential sorting algorithm, we use a very simple, non-recursive bottom-up variant of merge sort. For simplicity, we assume N, the number of keys to be sorted, a power of 2. The keys are stored in an array of length N. Merge sort works as follows: Assume that the array consists of $N/2^i$ neighboring sorted subsequences of length 2^i. Every pair of neighboring subsequences is merged to form a sorted sequence of length 2^{i+1}. For merging one pair, $2^{i+1} - 1$ comparisons are needed (worst case). Now $N/2^{i+1}$ sorted sequences are left. To sort the keys, this procedure is repeated for $i \in \{\log k, \ldots, \log N - 1\}$ where we assume that there are initially N/k sorted subsequences of length k (k also being a power of 2). Hence, the exact number of key comparisons is $\lambda_{MS}(N, k) = N \log(N/k) + 1 - N/k$.

The *latency* $L_{MS}(N, k, T_{SW})$ of a sequential implementation of merge sort is the time that goes by until the sequence of length N that consists of N/k sorted subsequences of length k becomes completely sorted, i.e.,

$$L_{MS}(N, k, T_{SW}) = (N \log(N/k) + 1 - N/k) \cdot T_{SW} \ , \tag{1}$$

where T_{SW} is an architecture constant depending on the clock period, cache size, the type of processor used, and overhead of the algorithm (loops etc.). For arbitrary sequences, $k = 1$. Hence, to completely sort an arbitrary sequence, $L_{MS}(N, 1, T_{SW}) = (N \log N + 1 - N) \cdot T_{SW}$.

2.2   Systolic Insertion Sort

The systolic insertion sort of Parhami and Kwai [5] belongs to a class of algorithms called *piecewise regular algorithms* (see [7, 8, 6]) Using the notation introduced in [6], the systolic sorting algorithm can be described in the following equations that define the data flow through a linear array of processing elements (Fig. 1 shows the physical structure of a processing element (PE)):

$$
\begin{aligned}
comp1[t,p] &= (a[t-1,p-1] \ge b[t-1,p]); \\
x1[t,p] &= \text{switch}((comp1[t,p]=1), b[t-1,p], \ (comp1[t,p]=0), a[t-1,p-1]); \\
x2[t,p] &= \text{switch}((comp1[t,p]=0), b[t-1,p], \ (comp1[t,p]=1), a[t-1,p-1]); \\
comp2[t,p] &= (x2[t,p] \ge c[t-1,p]); \\
a[t,p] &= \text{switch}((comp2[t,p]=1), x2[t,p], \ (comp2[t,p]=0), c[t-1,p]); \\
x3[t,p] &= \text{switch}((comp2[t,p]=0), x2[t,p], \ (comp2[t,p]=1), c[t-1,p]); \\
mux1[t,p] &= \text{test}(a[t-1,p-1],16); \\
c[t,p] &= \text{switch}((mux1[t,p]=1), x4[t,p], \ (mux1[t,p]=0), x3[t,p]); \\
mux2[t,p] &= \text{test}(c[t-1,p],16); \\
x5[t,p] &= \text{switch}((mux2[t,p]=1), b[t-1,p+1], \ (mux2[t,p]=0), c[t-1,p]); \\
b[t,p] &= \text{switch}((mux1[t,p]=1), x5[t,p], \ (mux1[t,p]=0), x1[t,p]);
\end{aligned}
$$

Here, 'switch$(c1,a1,c2,a2)$' is similar to an if-statement and evaluates $a1$ if $c1$ is true and $a2$ if $c2$ is true. 'test$(a1,b)$' returns the value of bit b of the control vector $a1$. [2] Furthermore, p is the processor index and t the time index.

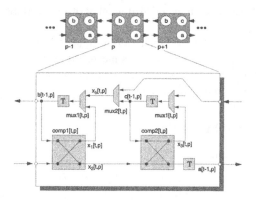

Figure 1: Detailed view of data flow in a single PE. The elements T denote the registers a, b and c highlighted above.

This regular algorithm can be interpreted as the computation of a linear systolic array of $P = N/2$ identical PEs. An example of such a computation for $P = 2$ is given in Fig. 2. At time steps $0,\ldots,N-1$ the unsorted sequence of length N is input serially into the leftmost PE. Each PE

---

[2]Bit 16 is a special tag bit here, refer to [5].

performs a compare/exchange operation on its input keys (where x denotes $+\infty$), which results in the serial output of the sorted sequence at time steps $N, \ldots, 2N-1$.

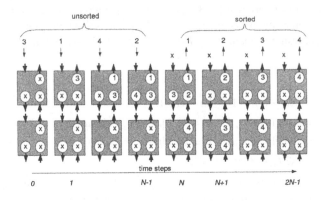

Figure 2: An example of the systolic sorting algorithm on a 2-cell linear array.

The latency of the systolic insertion sort algorithm is

$$L_{SIS}(N, T_{HW}) = 2 \cdot N \cdot T_{HW} \ .$$

(with T_{HW} the clock period of every PE.) When implementing the PEs on a FPGA chip co-processor, P is limited due to bounded area resources (see Sec. 5).

3 Combining the Algorithms

In this section, we first suppose an unbounded number of PEs, and then we assume a limited number P of PEs on the co-processor.

The runtimes are closely related to the ratio $\rho = T_{HW}/T_{SW}$ of the hardware speed and the speed of the software solution. Loosely speaking, this means the co-processor can execute $1/\rho$ key comparisons in the same time the uniprocessor can execute a single comparison. If the co-processor is faster than the processor, then $\rho < 1$.

3.1 Unbounded Number of PEs

3.1.1 Sorting a Single Sequence

First, we assume that for sorting a sequence of N keys, $P = N/2$ PEs are provided on the coprocessor such that entire sequence can be sorted by the co-processor exclusively. Furthermore, we assume the co-processors to be slow, i.e., $\rho > 1$. (Otherwise, the best runtime would obviously be achieved by using solely the co-processor.)

The hybrid algorithm works as follows: Initially, the sequence is partitioned into N/k subsequences, each consisting of k keys. These subsequences are transferred to the co-processor where $2P/k$ arrays of PEs of length $k/2$ sort each sequence using the systolic insertion sort algorithm. Then the sorted subsequences are returned to the sequential processor and finally merged using the merge sort algorithm. Fig. 3 gives a picture of how the hybrid algorithm works.

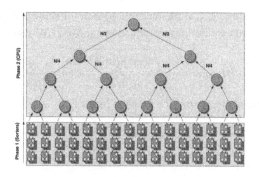

Figure 3: Scheme for hybrid sorting.

In the following, we compute the latency $L_{\text{hyb1}}(N,k,T_{\text{HW}},T_{\text{SW}})$ of this algorithm and determine k_{opt} such that $L_{\text{hyb1}}(N,k_{\text{opt}},T_{\text{HW}},T_{\text{SW}})$ is minimal. We have

$$\begin{aligned}
L_{\text{hyb1}}(N,k,T_{\text{HW}},T_{\text{SW}}) &= L_{\text{SIS}}(k,T_{\text{HW}}) + L_{\text{I}} + L_{\text{MS}}(N,k,T_{\text{SW}}) \\
&= 2kT_{\text{HW}} + L_{\text{I}} + \left(N\log(\tfrac{N}{k}) + 1 - \tfrac{N}{k}\right) \cdot T_{\text{SW}}
\end{aligned}$$

where L_{I} is the interface latency, i. e., additional time that is necessary for communication between processor and co-processor and for scheduling the N/k sequences. Let $\lambda_{\text{hyb1}}(N,k,\rho) = (L_{\text{hyb1}}(N,k,T_{\text{HW}},T_{\text{SW}}) - L_{\text{I}})/T_{\text{SW}}$, i. e.,

$$\lambda_{\text{hyb1}}(N,k,\rho) = N\log(\tfrac{N}{k}) + 1 - \tfrac{N}{k} + 2k\rho \tag{2}$$

That means that for simplicity, we assume $L_{\text{I}}/T_{\text{SW}}$ close to 0. $\lambda_{\text{hyb1}}(N,k,\rho)$ is roughly the number of sequential comparisons plus the number of 'PE comparisons' of the parallel algorithm.

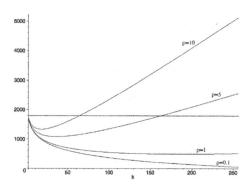

Figure 4: $\lambda_{\text{hyb1}}(256,k,\rho)$ for $\rho \in \{0.1,1,5,10\}$. The horizontal line is $\lambda_{\text{MS}}(256,1)$.

The λ_{hyb1}-curves of the hybrid algorithms (shown in Fig. 4) are below the merge sort line for a specific range of k. It can be seen that for $\rho = 10$, the hybrid solution is faster than the sequential one for values $k \le 64$, so the sequence should be devided into 4 subsequences at least. For $\rho = 5$, a speedup is obtained for $k \leq 162$, while the curves for $\rho \leq 1$ show that the sequence should be sorted by the systolic sorter exclusively. Minimizing $\lambda_{\text{hyb1}}(N,k,\rho)$ yields

$$k_{\text{opt}}(N,\rho) = \frac{N + \sqrt{N^2 - 8(\ln 2)^2 \cdot N\rho}}{4\rho \ln 2} \approx \frac{N}{2\rho \ln 2} \ . \tag{3}$$

Compared with sequential merge sort, the speedup factor is

$$S_{\text{hyb1}}(N,k,\rho) = \frac{\lambda_{\text{MS}}(N,1)}{\lambda_{\text{hyb1}}(N,k,\rho)} \ .$$

Fig. 5 shows the maximum speedup curve over N using $k_{\text{opt}}(N,\rho)$ as computed in Eq. (3). The number of keys sorted per time unit is defined as *data rate* ϑ_1:

$$
\begin{aligned}
\vartheta_1(N,k,T_{\text{HW}},T_{\text{SW}}) &= \frac{N}{L_{\text{hyb1}}(N,k,T_{\text{HW}},T_{\text{SW}})} \\
&= \frac{N}{L_{\text{SIS}}(k,T_{\text{HW}}) + L_{\text{I}} + L_{\text{MS}}(N,k,T_{\text{SW}})} \ .
\end{aligned}
$$

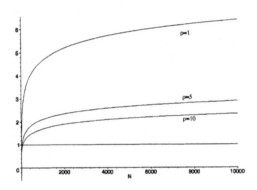

Figure 5: Speedup $S_{\text{hyb1}}(N, k_{\text{opt}}(N,\rho), \rho)$ for $\rho = \{1,5,10\}$.

3.1.2 Sorting Several Sequences Using Pipelining

When solving several individual equal-sized sorting problems subsequently, the co-processor does not need rest idle after having finished sorting a sequence. It can start sorting the next one. In Fig. 6, a schematic view of the resulting schedule is presented.

As the processor can still work on the merge sort while the co-processor is communicating, the interface latency is only associated with the latency of the systolic sorter. Now, the data rate is

$$\vartheta_2(N,k,T_{\text{HW}},T_{\text{SW}}) = \frac{N}{\max(L_{\text{SIS}}(k,T_{\text{HW}}) + L_{\text{I}},\ L_{\text{MS}}(N,k,T_{\text{SW}}))} \ ,$$

which is obviously maximized for $L_{\text{SIS}}(k,T_{\text{HW}}) + L_{\text{I}} = L_{\text{MS}}(N,k,T_{\text{SW}})$. Without proof, we note that this maximum is attained for $k \approx k_{\text{opt}}$ (taking L_{I} not into account).

Due to smaller idle times, a greater utilization of both the processor and co-processor occurs when processing in pipelining mode. As for the data rate, also the utilization is optimal for $L_{\text{SIS}}(k,T_{\text{HW}}) + L_{\text{I}} = L_{\text{MS}}(N,k,T_{\text{SW}})$,

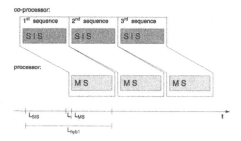

Figure 6: Pipelined sorting of 3 independent sequences.

3.2 Bounded Number of Systolic PEs

In the following, we examine speedup and cost trade-offs under the (more realistic) assumption of a limited number P of PEs that can be realized on the co-processor.

Again, the hybrid algorithm works in several phases. In the first phase, the sequence is partitioned in N/m subsequences with $m = 2 \cdot P$. The subsequences are sent to the co-processor where they are sorted by applying the first hybrid sorting algorithm presented in Sec. 3.1 using the pipelined approach similar to that described in Subsec. 3.1.2. In the second phase, the sorting is finished by applying merge sort on the remaining sequence.

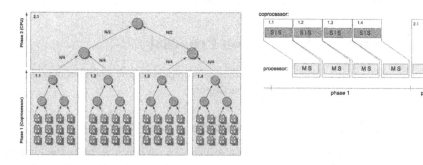

Figure 7: On the left, the scheme for hybrid sorting with a limited number P of PEs. Note that the leaf boxes (1.1,...,1.4) are processed by the co-processor sequentially. On the right, a schedule with $N/m = 4$.

Fig. 7 shows an example with $P = 12$ PEs that are partitioned in 4 linear arrays of 3 PEs each. In Phase 1, $N/m = 4$ subsequences each of length $2P = 24$ keys is pre-sorted. When the processor starts merging this subsequence, the co-processor can start the pre-sorting phase for the next subsequence. Finally, the 4 subsequences are merged in phase 2.

The latency of this schedule is

$$L_{\text{hyb2}}(N, m, k, T_{\text{SW}}, T_{\text{HW}}) = L_{\text{SIS}}(k, T_{\text{HW}}) + L_{\text{I}} + (\tfrac{N}{m} - 1) \cdot \max(L_{\text{SIS}}(k, T_{\text{HW}}) + L_{\text{I}}, \, L_{\text{MS}}(m, k, T_{\text{SW}}))$$
$$+ L_{\text{MS}}(m, k, T_{\text{SW}}) + L_{\text{MS}}(N, m, T_{\text{SW}}).$$

The latency of Phase 1 is minimized if $L_{\text{SIS}}(k, T_{\text{HW}}) + L_{\text{I}} = L_{\text{MS}}(m, k, T_{\text{SW}})$. Then, the equation simplifies to $L_{\text{hyb2}}(N, m, k, T_{\text{SW}}) = (\tfrac{N}{m} + 1) \cdot L_{\text{MS}}(m, k, T_{\text{SW}}) + L_{\text{MS}}(N, m, T_{\text{SW}})$.

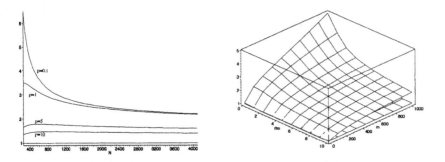

Figure 8: Maximal speedup over N for constant number P of PEs (left) and speedup over $m = P/2$ and ρ for constant total $N = 2048$ of keys (right).

Fig. 8 (left) shows the speedup curve over N for $P = 128$ PEs realized by the co-processor for $\rho = 0.1, 1, 5, 10$. That means that the length of the subsequences is $m = 256$. It can be seen that due to an increasing effort in the second phase of the hybrid algorithm, the speedup curves flatten for $N \gg P$. Fig. 8 (right) shows the speedup curve for a constant total of $N = 2048$ keys, in dependence of the number P of PEs and ρ. This graph can be seen as a generalized cost-speedup analysis curve because the number of PEs available on the FPGA chip relates to its price. The speedup increases with the number of PEs used until $P = 2 \cdot N$. If more PEs are available, the algorithm reduces to the algorithm of Sec. 3.1.

4   A Hardware Implementation Proposal

We propose an architecture consisting of a standard microprocessor and a number of co-processors for sorting, based on the shared memory paradigm.

4.1   A Shared Memory Based Architecture

Our architecture (shown in Fig. 9) is based on a system using at least one cache level in order to allow the processor and the co-processors to work in parallel. All data transfer between main memory and cache is performed in small blocks of appropriate size, generally a power of two. For this reason, the main memory consists of burst mode DRAMs.
We use s hardware sorting devices $S_1, \ldots, S_s$, each containing $g = P/s$ PEs (and thus able to sort a subsequence of $k = 2g$ keys) connected to the system bus via two FIFO buffer memories. A buffer memory is only necessary if the sorters cannot operate at the same clock rate as the burst-mode memory (e.g., due to clock rate limitations of the FPGAs). Additionally, some controller logic is required for address generation, FIFO and sorter control (AFSC).
We define the clock ratio c as the ratio between the clock rates of hardware sorters and CPU. To avoid synchronization problems, we assume $1/c$ an integer. In Fig. 10, it is shown how $s = 4$ sorting devices work in parallel for a clock ratio of $c = 2$. First, the input FIFO of the first sorter device (S_1) is loaded with a length k-sequence from the main memory. As soon as the input FIFO contains a key, S_1 starts operating. The loading process is finished after L_{trans}, and now the remaining input FIFOs are filled subsequently in the same manner. Ideally, when the last

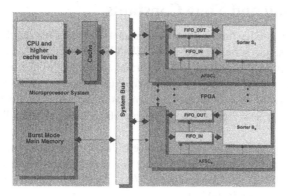

Figure 9: Shared memory based architecture.

sorter (S_4) starts working, S_1 has finished writing its output data to the output FIFO. Now, the transfer of the first sorted length k-sequence back to the memory is started, which takes L_{trans} again. This can not be done until a complete length k-sequence is in the output FIFO in order to avoid the FIFO running empty.

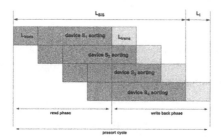

Figure 10: Schedule for $c = 2$ and $s = 4$.

4.2 Latencies and Optimal Number of Sorters

In order to exploit the maximum memory bandwidth during the hardware sorting process, an optimal number s_{opt} of sorting devices required for minimizing bus idle times will be derived. Assume $c = 1$. Let $L_{trans} = k \cdot T_{HW}$ be the time for transferring a length k-sequence from/to a FIFO and $L_{dev} = 2 \cdot k \cdot T_{HW}$ the time for sorting this length k-sequence in a sorting device. L_{dev} can be written as $L_{dev} = 2 \cdot L_{trans}$ since k clocks are used for transferring a length k-sequence and $2k$ clocks for sorting this sequence in a sorting device. With $c \leq 1$ this leads to $L_{dev} = 2c \cdot L_{trans}$, ($1/c \in \mathbb{N}$) resulting in $s_{opt} = \frac{L_{dev}}{L_{trans}} = 2 \cdot c$ as an optimal number of sorting devices.

5   Prototyping and Time Constants

In order to obtain a good estimation for the constant T_{HW} of a real hardware sorter device, we implemented a single sorter device using XILINX Virtex FPGAs on a rapid prototyping platform. The sorter device consists of 128 PEs arranged in a single chain S_1 ($s = 1$). We proved this configuration to work correctly at a clock rate of 66 MHz which leads to $T_{HW} = 15ns$. For determining the constant T_{SW} we took measurements of the real CPU time of the merge sort algorithm described in Sec. 2 on a Sun UltraSparc-60 (300 MHz) and a Pentium-II PC (300 MHz).

The merge sort was performed for each sequence of length N being a power of 2 in the range $64, \ldots, 524288$. The results for the UltraSparc are $T_{SW} = 0.29\mu s$ and $\rho = T_{HW}/T_{SW} = 0.052$ and for the Pentium-II PC $T_{SW} = 0.15\mu s$ and $\rho = 0.1$ (both for $T_{HW} = 15ns$). These numbers lead to speedup factors summarized in Table 1.

S	N			
	512	1024	2048	4096
Sun	3.96	3.59	3.08	2.67
PII	3.92	3.57	3.07	2.67

Table 1: Theoretical max. Speedup S over N for Sun and PII each with a single FPGA chip ($P = 128$ PEs). (We use $k = m/2$ here since k_{opt} does not exist in this example because $L_{SIS} < L_{MS}$).

Interestingly, we obtain spectacular speedup factors of up to 100 if more than one FPGA may support the microprocessor.[3] These are reported in Table 2 together with the number or required Xilinx FPGAs to obtain these speedups.

S	N			
	512	1024	2048	4096
Sun	77.4	87.0	96.7	106.3
PII	40.0	45.0	50.0	55.0
#FPGAs	2	4	8	16

Table 2: Max. Speedup S obtained over N for Sun UltraSparc-60 and Pentium-II PC when each supported by multiple FPGA chips of type Xilinx Virtex (XCV-1000-4) as co-processor.

6   Summary and Conclusions

In this paper, we have evaluated hybrid sorting algorithms that combine the classical sequential merge sort algorithm running on a microprocessor with a systolic insertion sort algorithm running on an FPGA-based co-processor.

We investigate the theoretical bounds for latencies, speedups and utilization for hybrid sorting and merging algorithms. We have also proposed and implemented (in part) a real architecture

---

[3]These maximal speedups correspond to co-processor-only sorting due to the measured values for $\rho < 1$.

using available technology (i.e., Pentium processor and Xilinx FPGA as co-processor) for which our measurements have proven the validity of our theoretical results.

In the future, it would be very interesting to investigate other combinations of well-known sorting algorithms for cost/speedup analysis tradeoffs. The potential of the proposed architecture seems to be high also for computation intensive algorithms from other application areas.

References

[1] K. E. Batcher. Sorting networks and their applications. In *AFIPS Conf. Proc. 32*, pages 307–314, 1968.

[2] D. E. Knuth. *The Art of Computer Programming, Volume 3: Sorting and Searching.* Addison-Wesley, Reading, Massachusetts, 2nd edition, 1998.

[3] M. R. Kramer and J. van Leeuwen. Systolische Berechnungen und VLSI. *Informatik Spektrum*, 7:154–165, 1984.

[4] C. E. Leiserson. Systolic priority queues. In *Proc. Conf. Very Large Scale Integration: Architecture, Design, Fabrication*, pages 199–214, 1979.

[5] B. Parhami and D.-M. Kwai. Data-driven control scheme for linear arrays: Application to a stable insertion sorter. *IEEE Transactions on Parallel and Distributed Systems*, 10:23–28, 1999.

[6] J. Teich. *A Compiler for Application-Specific Processor Arrays.* Shaker (Reihe Elektrotechnik). Zugl. Saarbrücken, Univ. Diss, ISBN 3-86111-701-0, Aachen, Germany, 1993.

[7] L. Thiele. On the design of piecewise regular processor arrays. *In Proc. IEEE Symp. on Circuits and Systems*, pages 2239–2242, 1989.

[8] L. Thiele. CAD for signal processing architectures. In *The State of the Art in Computer Systems and Software Engineering (P. Dewilde, ed.)*, Boston: Kluwer Academic Publishers, 1992.

HETEROGENEOUS SYSTEM DESIGN

System Approach and Methodology

COMPUTER AIDED DESIGN FOR MICROELECTROMECHANICAL SYSTEMS

J. Mehner, J. Wibbeler, F. Bennini and W. Dötzel
Department of Microsystems and Precision Engineering
Chemnitz University of Technology, D-09107 Chemnitz
mehner@infotech.tu-chemnitz.de

Abstract

With the rapid development in the field of microtechnologies there is a growing need for fast and accurate design tools. Latest microsystems which are widely used as sensors and actuators comprise complex geometrical structures, different transducing principles, advanced electronic circuitry and digital signal processing units. A comprehensive computer support is required to get a full understanding of the device behavior, to optimize the system performance and to enable competitive industries to get time and cost efficient prototypes.

The entire design process should capture conceptual assistance to find practical configurations, physical simulations at the components level, parameter reduction techniques (also called macromodeling), a system level description for the whole device and tools to optimize the fabrication process. This paper presents an overview of present design strategies, tools and algorithms which are applied successfully in microelectromechanical (MEMS) components design. Special emphasis is put on the simulation of multiple energy domains and macromodel extraction techniques.

1.   Introduction

Components design is one step of the whole microsystem's design procedure. In particular it deals with a special piece of the system that can be considered as a separate unit. In the following paper the mechanical part will be regarded as the component of our interest. Distinguishing mark of mechanical components is that at least one part moves against one other and that this motion is absolutely necessary for the operation. Mechanical components contain flexible regions (beams, membranes, plates) or rigid bodies rotate on bearings or move on a slider. This motion is used either to sense physical quantities (accelerations, forces, temperature changes, etc.) by transforming them into electrical signals or physical quantities are applied on purpose to drive parts of the structure to a desired position (e. g. microscanner, positioning systems). In both cases accompanying fields are necessary to transduce different physical energy domains. Those physical domains are directly associated with the mechanical part and need to be considered in our component models at once.

R. Merker and W. Schwarz (eds.), System Design Automation, 111-130.
© 2001 *Kluwer Academic Publishers.*

Other frequently used components in microsystems are for instance fluidic, optic, biologic, chemical or electronic ones. All of them are working together and form the systems functionality. Simulations at the system level are carried out to assess their cooperative play. The design process itself is iterative and interactive requests between the different levels are up-to-date ("meet in the middle strategy" [1]). The ultimate goal of microsystems design is that the assembled system fits best together and not that each single component gets best performance. Component design starts with specifications given by the system designer. Those specifications can be classified in:

- Demands on the mechanics (compliance in different directions, inertial mass, natural frequencies, fracture strength),
- Physical and geometrical restrictions (electric voltage, outer frame size, optical area),
- Available or preferred technology (bulk or surface technology, SCREAM, LIGA),
- Operating conditions (vibration amplitudes, temperature cycles, mounting angle),
- Environmental conditions (temperature, moisture, pressure, dust, aggressive mediums).

The component design procedure is governed by two basic objectives. First a structure has to be found that fulfills all above requirements and second the structure's behavior should be described properly. This task is complicated by the fact that different physical phenomena are acting on the same part of a structure simultaneously. For example, the membrane of a pressure sensor is flexure and mass in the mechanical domain, part of the electrostatic system in the electric domain and moving wall of a squeezed gap in the fluidic domain. Furthermore parasitic phenomena like intrinsic membrane stress, thermal mismatch of different materials and thermal noise caused by air molecule collisions on the membrane should call our attention. The component designer must have a full understanding of all involved physical disciplines, in numerical mathematics, computer science and microtechnologies which is a big challenge.

What are the goals of component design? Following the design procedure it turns out that several shape elements would fulfill all specifications in a similar manner. Mostly, each of them can be optimized according to a mathematical criterion, but hardly one can decide which shape element would really be the best. Doing a promising selection depends a lot on the experience of the designer. Nevertheless, one should have in mind that a microstructure is only qualified for a given application if all of the following issues are accomplished:

1. All system specifications are in a permissible range,
2. Function is guaranteed even at rough but admissible operating and environmental conditions,
3. The component has sufficient fracture strength and lifetime according to the task,
4. Dynamic loads like vibrations or jerks do not disturb the performance (no internal resonance),
5. Manufacturing tolerances do not diminish accuracy and yield bounds,
6. The information, energy and medium exchange between component and device is solved properly,
7. The component can be fabricated time and cost efficient with the available equipment.

A number of simulation tools have been applied in the past for the design of mechanical components. Their theoretical approach, known as continuum mechanics, is still applicable despite of the immense miniaturization of all geometrical dimensions in microsystems. In [2] was shown that even structures with thickness down to 10 nm behave like macroscopic ones. Nevertheless, specific problems arise from applied technology, used materials, advanced operation schema and growing integration density. In microsystems technology strong restrictions appear in terms of shape design. It can only be design what technology is able to manufacture and this is commonly very limited. Furthermore, the basis material single crystal silicon behaves anisotropically and singular stress concentrations occur at etched V-notches. Strong aspect ratios of geometrical dimensions, bi-directional interactions of physical domains and coupling of field simulators to the electronic design environment are difficult to handle. It has clearly been enormous progress in developing CAD tools for microsystems during the last years but there are still a lot of points at issue.

2.   Design procedure of micromechanical components

Today the design procedure in digital microelectronics is almost completely computer guided. Such a high level of computer support became possible since the entire system is composed by standard elements which are available in model libraries. Consequently, using advanced simulation tools and proper hardware description languages the electronic design can be done nearly automatically. Microelectronic design starts from the most general case, the algorithmic level and leads to a detailed chip layout. This methodology is known as *top-down strategy*.

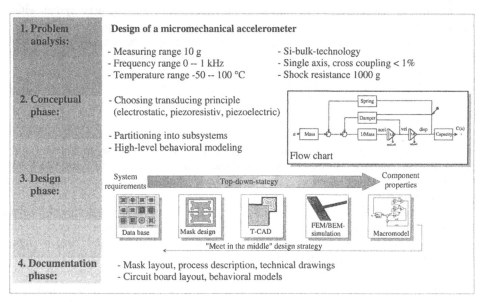

Fig. 1:   Exemplary design flow of micromechanical components

Looking at commercial design tool in micromechanics (ANSYS, MEMCAD, MEMS/Pro) there is a completely different approach. Their design procedure starts from the mask layout, is followed by numerous physical simulations and finally an approximated macro-model is generated which can be used at the system level. This technique is called *bottom-up strategy*. In practice, of course, the designer is not able to start at the bottom level because he doesn't know what layout is suitable. Desirable in mechanics are model libraries of frequently used shape elements which can be mounted together to obtain the systems functionality. This approach would allow the same procedure as known from microelectronics. First steps in this direction are published. Yet, some libraries are available for simple design tasks like accelerometer or pressure sensor design [3], [4]. Those libraries are very promising but the trouble is that mechanical components are much more manifold than electronic ones. It will take a longer period to fill those libraries with eligible elements for any arbitrary application. For this reason a combination of both strategies is presently suggested. The component designer tries to find a qualified candidate in the model library and goes through a loop of design cycles to find a proper mask layout. Backwards he simulates the physical behavior by numerical techniques accurately and establishes a macro-model. This strategy is called *"meet in the middle"* in literature [1].

3.   Computer aids for the conceptual design

Each engineering design procedure starts with a comprehensive problem analysis where all specifications are checked concerning discrepancies and incompleteness. By now this step can be done computer aided. Lets illustrate this statement by the previous accelerometer design procedure of Fig. 1. Measuring range and bandwidth are 10 g and 1 kHz respectively. The mechanical component of such an accelerometer is usually a spring mass system with a damping ratio of about 0.7 and a natural frequency f which is at least three times higher than the upper frequency bound f_u. Applying simple rules of physics one can assess a deflection limit u_n at 10 g by:

$$3f_u < \frac{1}{2\pi}\sqrt{\frac{c}{m}} = \frac{1}{2\pi}\sqrt{\frac{a_n}{u_n}}$$

$$u_n < \frac{a_n}{(2\pi\,3\,f_u)^2} = 0.3\ \mu m \qquad (1\text{-}2)$$

where a_n is the maximum acceleration, c the spring stiffness and m the seismic mass. Eq. (1) shows that the transducer electronic circuit must resolve a displacement of 0.3 μm which may be hard to realize. This deflection limit is independent of the chosen shape element or any geometrical dimensions.

During the conceptual phase most microstructures can be described by one degree of freedom systems but this is by far not necessary. At a high abstraction level the behavior of single elements is reasonable modeled by black-boxes and their interactions are described by signal junctions and mathematical operators. A data base of signal flow charts should be established to compare several transducer principles quickly. The component designer himself decides how detailed each element needs to be characterized. In our case the spring force is given by a bilinear function to consider the stop position of the seismic mass at the ground electrode. In contrast damping behavior is modeled more precisely assuming squeeze film theory and a squared capacitor plate. Fig. 2 illustrates how a flow chart may look like for most bulk micromachined accelerometers.

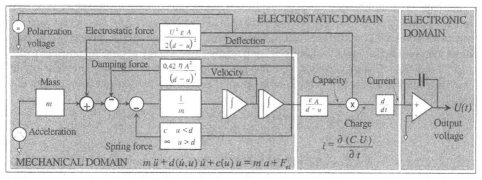

Fig. 2: High-level behavioral model of a capacitive accelerometer

Signal flow charts are the link between general specifications and shape design parameters (phase three in Fig. 1). At this point, the designer knows what stiffness the flexure should have, what electrode size and gap is proper as well as optimal data for other quantities. The model above estimates not only the static response but also transient properties of the future system like settling time and detectable acceleration pulse length. Furthermore, and this is very important, one gets information about the influence of manufacturing tolerances to the output signal. Yield, accuracy, signal to noise ratio and drift computations become possible if statistical process parameters are considered. Unfortunately, there is often a lag on information.

4. Physical modeling of microelectromechanical components

In the following step the designer seeks for a geometrical structure with appropriate properties as calculated in the conceptual phase. A straightforward approach for structural synthesis is by now impossible. Presently computer aided synthesis in mechanics is very limited. Examples are known from transmission gearing design or is applied to special parts of a structure like airplane wings. Capturing the whole manifold of three dimensional space would exceed computational capabilities by far. In practice, computer support is confined to proposals for proper shape elements but those have to be implemented in libraries. However if such a shape element is eligible for a given task there are powerful toolboxes available for dimensional design and shape optimization.

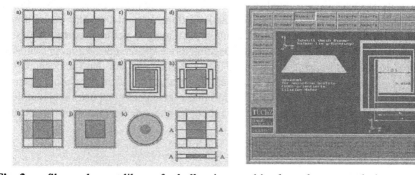

Fig. 3: Shape element library for bulk-micromachined accelerometer design

Some shape element libraries are developed for specific applications where spring mass systems are sufficient. One example is shown in Fig. 3. This library contains frequently used microstructures for bulk or surface micromaching which are successfully applied for accelerometer design. Each element is described analytically by an energy approach known as Castigliano's principle. Energy methods are convenient if different types of work are done on the same structure simultaneously. Eq. (3) adds up stretch, bending, torsion and shear energy of a beam

$$W = \int \frac{F_n^2(s)}{2E_xA}ds + \int \frac{M_b^2(s)}{2E_xI}ds + \int \frac{M_t^2(s)}{2GI_t}ds + \int \chi \frac{F_Q^2(s)}{2GA}ds \qquad (3)$$

were W is the total energy, F_n the normal force, M_b the bending moment, M_t the torsion moment, F_q the shear force, E_x Young's modulus, G the shear modulus, I the moment of inertia, I_t the torsional moment of inertia, A the cross section area, χ the shear constant and s the integration path along the beam axis. This equation can be extended by non-mechanical energy terms too (e.g. electrostatic ones). Calculating the derivatives to the acting force one obtains the caused deflection at any arbitrary structural point. The very benefit of this method is that finally a mathematical term exists for all component properties rather than a number. Any behavioral simulation is directly related to the applied material properties and to all geometrical dimensions.

Furthermore, sensitivity analysis and shape optimization become simple to handle but unfortunately resolving the energy term of eq. (2) sometimes turns out to be difficult.

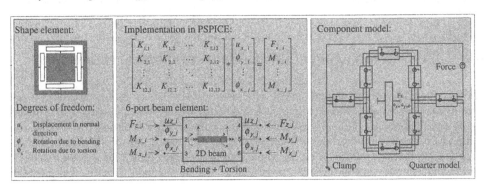

Fig. 4: Direct implementation of beam elements in PSPICE

Another very promising approach can be applied if the mechanical system consists of rigid bodies which are suspended by a number of straight beam elements. Numerically efficient matrix representations for the interface behavior (deflection force relationship) can be found in textbooks. A similar description is used in circuit simulators to define n-port elements. In the same way as non-mechanical elements are assembled to a system our beam representation can be applied to model mechanical systems (Fig. 4). Advantages are that the designer can place his own beam arrangements (even non-linear ones), that the electronic circuitry can directly be coupled (due to the use of PSPICE) and that a graphical composition is simple to handle and apparent. Drawback is that a mere number and not a mathematical term exists as a solution of the problem.

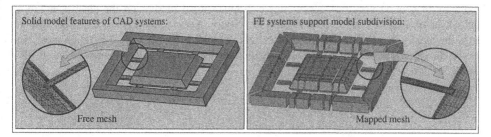

Fig. 5:   Different discretization techniques used in solid modeling tools

The third group of simulation techniques is based on a 3-dimensional solid model of the microstructure. Solid models can either be a result of a designer's work using CAD software tools or they are generated automatically from the mask layout by etching simulators. In both cases one will obtain a complicated volume represented by many boundary areas. Commercial simulation tools apply discretization techniques as the finite element method (FEM) to analyze their behavior. Accuracy, claimed computer resources and CPU-time depends mainly on the quality and quantity of the employed FE-mesh. The trouble is that presently all commercial mesh generators produce exclusively tetrahedral elements if there are more than six boundary faces. Tetrahedral meshes make the computation cumbersome since parabolic shape functions are necessary (more nodes) and throughout a higher mesh density is created (more elements). Alternatively, a manual subdivision of the solid model should be preferred. It is much more time consuming to establish a hexahedron (mapped) mesh but if such a subdivision is generated once it can be used efficiently several times.

Finite element computations allow a very precise prediction of the mechanical behavior in a reasonable time (usually minutes). Considering non-linearities, estimation of fracture strength, fatigue analysis, contact problems, friction and buckling phenomena became state of the art. Especially in terms of microsystems, the most important progress is that commercial tools (ANSYS, MEMCAD, ADINA) support multiple energy domain simulations.

5.   Multiple energy domain simulations

5.1   Interactions in electromechanical systems

Most transducers in microsystems technology are based on electrostatic fields. Reasons for this popularity are that electrostatic arrangements are simple to manufacture, a large field density becomes possible within small gaps, sensor and actuator effect may be in the same space, temperature drift is irrelevant and no additional mass is required (e.g. permanent magnets). Electrostatic fields around electrodes are governed by Poisson's differential equation. Once their solution is known the designer can calculate specific values as charge density, capacities between electrodes and infinity, electrostatic forces and so on. Therefore three techniques are applied mostly. First, an analytical approach is promising in case of small gaps between electrodes. If so, electrostatic field density depends simply on the local separation $d(x,y)$ and can be calculated analytically (Fig. 6). Second, finite element tools provide access to Poisson equation via thermal or magnetic analogy relationship. This method is widely used since a direct coupling to the mechanical part becomes possible. Third, boundary element tools are in general very fast, since simple 2-dimensional meshes on the electrodes are required. In general they capture the field to the infinity much better and consequently their results are more precise. Drawbacks occur in case of small gaps. There is still a high mesh density required despite of the quite homogenous field within. Presently a combination of FEM and BEM appears opportune (Fig. 6 right).

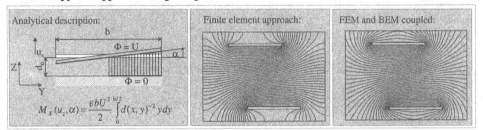

Fig. 6: Methods to analyze electrostatic fields, capacities and forces

All electromechanical systems are bi-directionally coupled. Any change of the electrostatic field density affects the mechanical system by altering the Maxwell forces which are acting on electrodes or dielectric layers. Reverse, any change of the mechanical position leads to a change of electrostatic properties like capacities. Criterion for a stationary position of an electromechanical system is a force equilibrium at all degrees of freedom. This shall be illustrated by two frequently used actuator principles, a single and a differential capacitor arrangement (Fig. 7). Their behavior gets obvious by studying the force-deflection function at a point placed on the seismic mass. In general, the remaining force F on this point at any deflection u is the sum of external mechanical forces (e.g. gravitation), electrostatic forces and spring forces. The remaining force F drives the system either to a equilibrium position where this force gets zero or to a stop position at the ground electrode. Usually there occur multiple equilibrium states, as shown in Fig. 7. Only one of them is a stable equilibrium state, the others are unstable positions, despite of the force balance. Criterion for a stable equilibrium is a negative slope of the force deflection function at this point. In practice, one has to compute the first derivative of F with respect to the deflection u to decide whether this state is stable or not. Furthermore it can be shown in Fig. 7 that any perturbation (jerks, vibrations) will only be compensated if the seismic mass remains inside the stable region bounds. Consequently, the components designer has to watch the behavior in the entire design space, he has to be careful in choosing initial values for iterations and to distinguish unstable equilibrium states.

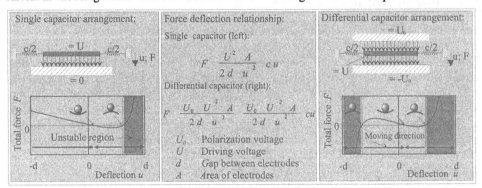

Fig. 7: Stable and unstable equilibrium positions of electromechanical systems

Studying the force deflection relationship of both systems (Fig. 7) it turns out that the single capacitor arrangement above deflects quadratically with the applied voltage while the differential capacitor changes linearly for small deflections. Furthermore the second arrangement requires a minimal spring stiffness which depends on the polarization voltage U_0. A very important design quantity for most actuators is the deflection range. Limits result either from the breakthrough voltage in air or the well known pull-in effect of electromechanical systems. In general transversally driven systems can only be deflected as high as one third of the gap to the ground electrodes. As soon as the deflection exceeds u_{PI} (deflection at pull-in) there is a lower driving voltage required than applied and consequently the seismic mass snaps to the ground. In case of differential capacitors this limit lies usually lower and depends mainly on the ratio c/U_0.

Remarkable on electromechanical systems is that electrostatic fields may alter mechanical properties. This effect can be used on purpose to calibrate electrostatic systems, to overcome manufacturing tolerances or to alter the system parameters during operation [5]. A Taylor series representation of the deflection dependent electrostatic force F_{el} shows clearly the influence of the applied voltage U to the spring stiffness c in case of small motions ($u<<d$)

$$m\,\ddot{u}+d\,\dot{u}+c\,u = \frac{U^2\varepsilon A}{2(d_{AP}-u)^2} = F_{el}(u)+\frac{\partial F_{el}}{\partial u}u+\cdots \qquad (4)$$

Calibration of mechanical systems by electrostatic forces makes sense in case of differential capacitor arrangements. Here two voltages are given to control the mechanical system: a polarization voltage U_0 and a driving voltage U. The mechanical stiffness and the deflection of the seismic mass itself can be altered independently by U_0 and U, respectively (see Table 1). A summary of electrostatic tuning capabilities and applications is given in [6].

Table 1: Characteristic data of 1-D transversely driven electrostatic actuators

	Single capacitor arrangement	Differential capacitor arrangement
Voltage-deflection function ($u<<d$)	$u = \dfrac{\varepsilon A}{2\,c\,d^2}U^2$	$u = \dfrac{2\,U_0\,\varepsilon A}{c\,d^2}U$
Stiffness requirement	$c>0$	$c > \dfrac{2\,U_0^2\,\varepsilon A}{d^3}$
Pull-in effect (U_{PI} pull-in voltage)	$U_{PI} = \sqrt{\dfrac{8\,c\,d^3}{27\,\varepsilon A}}\qquad x_{PI}=\dfrac{d}{3}$	$u_{PI} \le \dfrac{d}{3}$
Resonance frequency	$f = \dfrac{1}{2\pi}\sqrt{\dfrac{c-\dfrac{U^2\varepsilon A}{d^3}}{m}}$	$f = \dfrac{1}{2\pi}\sqrt{\dfrac{c-\dfrac{2\,U_0^2\,\varepsilon A}{d^3}}{m}}$

In the previous example a linearization of the governing differential equation of motion was applied. This common engineering approach is valid if small oscillation amplitudes at the operating point occur. The operating point itself may undergo large displacements. Otherwise the governing equation of motion becomes non-linear and numerical techniques are obligatory. The left graph in Fig. 8 shows capacity time functions of a harmonically oscillating mass at different motion amplitudes. Deviations from a sinusoid response increase as amplitudes are growing. On the right one can observe a significant difference of the natural frequency at the "high" and "low" voltage level.

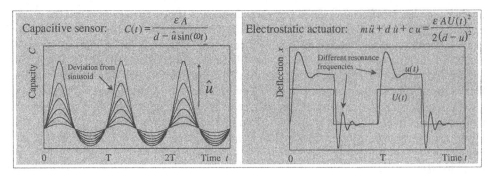

Fig. 8: Non-linear effects of electromechanical systems at large deflections

5.2 Simulation of multiple energy domains at the physical level

Simulations where the interactions of two or more disciplines of engineering are considered are known as *coupled-field analysis*. Coupled fields are classified by Zienkiewicz [7] in uni- and bi-directionally coupled problems. Uni-directionally coupled systems are simple to handle since the problem A can be analyzed separately and its results are taken into account in a subsequent analysis of problem B (thermal-stress analysis). The procedure for bi-directionally coupled systems depends on the region where in particular the interactions occur. If both fields are involved in the same domain and they interfere each other in the same space then the constitutive law should be used to establish a finite element description containing the degrees of freedom of both fields (matrix coupling). An example is a piezoelectric transducer. In contrast, if the fields are separated in a different space and the interactions occur at the interface between both then a load vector coupling arises more convenient. Examples are electrostatic transducers and fluid flow – structural analysis.

Table 2: Techniques to solve coupled fields by the finite element method

	Matrix coupling -piezoelectric effect-	**Load vector coupling** -electrostatic structure coupling-
Constitutive law:	Problem A: $\sigma = C\,\varepsilon - e\,E$ Problem B: $D = e^T\,\varepsilon + \in E$	Problem A: $\sigma = C\,\varepsilon$ Problem B: $D = \in E$
Finite-Element-approach:	$\begin{bmatrix} K_m & K_k \\ K_k^T & K_e \end{bmatrix} * \begin{bmatrix} u \\ \varphi \end{bmatrix} = \begin{bmatrix} F_m \\ Q \end{bmatrix}$	$\begin{bmatrix} K_m & 0 \\ 0 & K_e \end{bmatrix} * \begin{bmatrix} u \\ \varphi \end{bmatrix} = \begin{bmatrix} F_m + F_{el} \\ Q \end{bmatrix}$
Couple term:	$K_k = \int B_u^T e B_\varphi\,dV$	$F_{el} = \int N_u^T T_{el}\,n\,dA$
Characteristics:	• Single iteration (linear case) • Wavefront increased (time consuming) • Newton approach applicable (excellent convergence behavior) • User element necessary	• Multiple iterations (even in linear case) • Efficient due to partitioning • Relaxation approach (convergence only at weak coupling) • User programmable feature

The approach of multiple energy domain simulations will be discussed in detail for electrostatic transducers. Essential interactions are shown in Fig. 9.

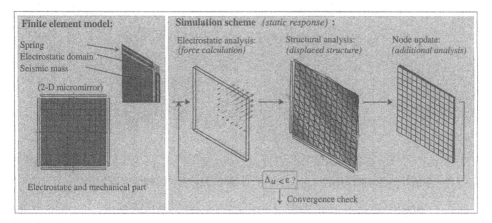

Fig. 9:   Simulation scheme of electromechanical systems using ANSYS (static case)

Electrostatic fields act on the mechanical part by electrostatic forces on electrodes or dielectric surfaces. Hence an ideal case of load vector coupling exists. Nevertheless it should be
mentioned that the influence of electrostatic fields on the movable part may also be described
by the internal stress state in air according Faraday's law. Similar to the piezoelectric effect,
electrostatic transducers can be captured by a (now non-linear) matrix coupling [8].

Electrostatic forces F_{el} at the movable part and capacities C are extracted from the results of a
static finite element simulation. Both values are calculated at the surfaces of electrodes or dielectric layers by

$$F_{el} = \int \varepsilon\, \vec{E}\left(\vec{E}\, \vec{n}\right) - 0.5\, \varepsilon\, \vec{E}^2 \vec{n}\; dA \tag{5}$$

$$C_{ij} = \frac{\sum Q_i}{U_j} \tag{6}$$

where ε is the permittivity in air, E the electric field density vector, Q the charge on the electrodes, n the unit normal vector, A the surface area and U the applied voltage. Reverse, the
mechanical system acts on the electrostatic part due to a shift of electrodes which thereafter
changes electrostatic properties. This shift is considered at the electrostatic domain by a node
update. Applying boundary element techniques for solving electrostatic fields one has to add
the deflection to the node coordinate, in case of finite element techniques usually all nodes in
the whole region have to be modified. Therefore a separate finite element run is necessary as
shown on the right in Fig. 9. Alternatively, some authors modify Young's modulus of the
electrostatic domain to a very small value ($E \rightarrow 0$) and keep the mechanical domain and electrostatic domain attached. Then the air region follows automatically the motion of the movable part but it is proven that the feedback often leads to wrong results.

From the mathematical point of view the iteration scheme used in Fig. 10 is known as relaxation method. This approach leads to a convergent solution if the spectral radius is smaller
than one [9]. In terms of physical quantities convergence occur if the iteration starts in the
stable region and, of course, if a stable equilibrium exists at all. Convergence speed (number
of iterations) depends on the slope of spring and electrostatic force functions (Fig. 11). Weak
convergence emerges if both functions have almost the same slope, for instance close to pull-
in.

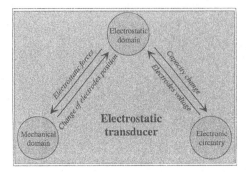

Fig. 10:   Interference scheme   Fig. 11:   Convergence scheme

Complete voltage deflection functions are desirable for microsystems design. A common approach is to apply the electrodes voltage, to compute the resulting deflection for a number of points and to do a non-linear function fit. It turns out that an inverse technique is much faster. Inverse means that a deflection is applied first and then the required voltage is determined. This approach can be accelerated essentially by a combination of analytical and numerical methods (see Fig. 12).

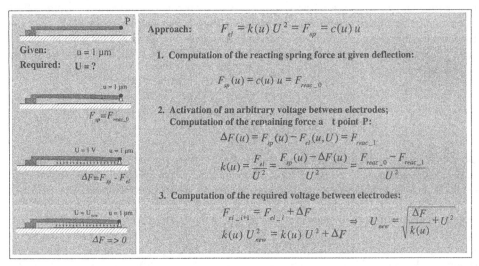

Fig. 12:   Simulation scheme for the inverse problem (given deflection → required voltage)

One approach is as follows. In equilibrium the electrostatic field must act with a same force as the suspension reacts at that position. The electrostatic force can be decomposed in a deflection dependent constant $k(u)$ and applied voltage squared U^2. Two finite element simulations are performed where the point of interest P is displaced to the desired position u. First analysis is done simply on the mechanical system. During the second analysis an arbitrary voltage must be applied in addition. The reacting force at the first run gives us the force the electrostatic field should have F_{reac_0} while the second run gives us the level of the force unbalance F_{reac_1}. Both values allow us to compute the constant $k(u)$ and finally the required voltage U_{new} for equilibrium analytically. State of the art finite element tools provide access to multiple energy domain simulations but the application engineer is enforced to exploit fast and robust algorithms as illustrated in this example [10].

5.3   Coupling of finite element tools and circuit simulators

Further demand on microsystems design is the capability to include electronic components into the simulation procedure. Especially for self test, self calibration and controller feedback design consistent models of both electronic circuitry and continuous components are required. Generally three methods are possible:

- A simulator coupling of finite element tools and electronic simulators (physical level)
- A macromodel approach which contains heterogeneous components (different levels)
- A transformation of all non-electrical components into the electric domain (network level)

The first method is very precise but mostly too cumbersome for everyday design tasks. Simulator coupling is used to investigate special problems as settling time, cross-coupling to adjacent cells of array structures and non-linear effects.

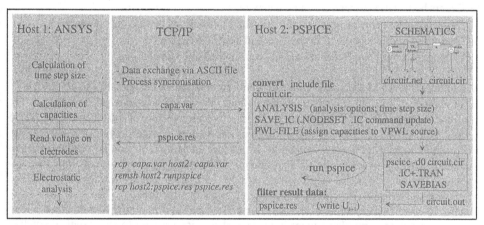

Fig. 13:   Schematic view on a simulator coupling of ANSYS and PSPICE

A simulator coupling scheme for transient behavioral analysis of electromechanical systems is shows in Fig. 13. The finite element tool ANSYS takes the part of continuous domain simulation and PSPICE takes the electronic part. Within each time step ANSYS calculates capacities between electrodes which are used later as an input quantity for a voltage controlled capacitor in PSPICE. After resuming all initial conditions from the previous time step PSPICE calculates the electrode voltages for the next ANSYS electrostatic domain simulation. File transfer and synchronization is done by the program "Parallel virtual machine" via computer network. This procedure is published in [11].

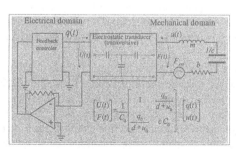

Fig. 14:   Equivalent circuit model [12]

Modeling at the network level is widely used to analyze electromechanical systems in SPICE. Unfortunately, a network representation is confined to small signal behavior (linearized equations) and usually applied for systems with a few degrees of freedom. On the other hand, those models provide a deep insight into the dynamic behavior of the transducer and they are fast and accurate. The constitutive matrix in Fig. 14 can be derived from the electrostatic energy function which is either a result of finite element calculations or obtained analytically. A very interesting approach to take benefit of both are macromodels.

6. Computer aided macromodeling for MEMS

Macromodels are defined as low-order behavioral representations of a device with the following attributes [13]:

- Expressed in a simple-to-use form, either as an equation, a network analogy or a small set of coupled ordinary differential equations,
- Represents suspended rigid bodies as well as flexible structures,
- Covers both stationary and dynamic behavior,
- Describes reasonably accurate the function and interactions of all components,
- Comprises multiple physical disciplines and electronic circuitry,
- Includes essential non-linear effects as occur in the large signal range.

Macromodels can be considered as a link between component design and system design. A large number of models are presented throughout the MEMS literature but mostly they are confined to suspended rigid bodies as the mass of accelerometers or gyroscopes [14]. Rigid bodies are completely described by six degrees of freedom. Further, only a few degrees are really important for the system operation. In practice most systems can be captured by two dominant motions which leads to an ordinary differential equation system with two unknowns. The major issue in creating macromodels is how to handle flexible structures like membranes, cantilevers, microbridges or thin micromirror structures.

Analyzing continuous systems which are governed by partial differential equations has a long history. For complex problems there is in general no direct access to an analytical solution available. A common engineering approach is to approximate the unknown solution T by a series of weighted shape functions

$$T(t,x,y,z) \approx \overline{T}(t,x,y,z) = \sum_{i=1}^{n} a_i(t)\, \phi_i(x,y,z) \tag{7}$$

where a_i are time dependent scale factors and ϕ_i are spatial functions. Substituting eq. (7) into an arbitrary partial differential equation (PDE) would lead to a system of n ordinary differential equations (ODE). Such an analytical solution procedure is well known as Galerkin method.

The following macromodel generation procedure is based on a numerical parameter extraction technique, since this allows to handle complex structures automatically. Now, the shape functions are not substituted into the PDE directly. Rather they are considered as globally defined degrees of freedom (generalized coordinates). Usually mechanical degrees of freedom are referred to a motion of a particular point, but shape functions describe a deformation state of a structure. The final deflection becomes the sum of spatial shape functions scaled by the factors a_i. From the theoretical point of view any set of shape functions can be applied as long as they don't violate the boundary conditions and they are linearly independent. Which shape functions would really fit best? One should have in mind that any deformation state of a flexible body in the static and dynamic case can be regarded as a superposition of eigenmodes. Choosing eigenmodes as shape function we receive a minimal number of equations in the final solution but the main point is that the shape functions can be computed automatically by a finite element run.

Next step is to establish the equations of motion. The ultimate goal is to consider not only mechanical loads but also forces which are related to other energy domains (e.g. electrostatic ones). Therefore the further model should base on energy terms since energy is independent from the physical meaning. The link from the energy formulation to the equations of motion is given by Lagrange's principle

$$\frac{d}{dt}\left(\frac{\partial E_k}{\partial \dot{a}_i}\right) + \frac{\partial E_p}{\partial a_i} = 0 \qquad i=1..n \tag{8}$$

where E_k is the kinetic energy and E_p the potential energy of the system. For an instantaneous deflection state both energy terms are defined as

$$E_k = \sum_{i=1}^{n}\sum_{j=1}^{n}\frac{1}{2}\left(\dot{a}_i\,\phi_i\right)^T M \left(\dot{a}_j\,\phi_j\right) \tag{9}$$

$$E_p = \sum_{i=1}^{n}\sum_{j=1}^{n}\frac{1}{2}\left(a_i\,\phi_i\right)^T K \left(a_j\,\phi_j\right) \tag{10}$$

where M if the finite element mass matrix and K the finite element stiffness matrix. Substituting eqn. (9) and (10) into the Lagrangian equation, we obtain for the linear case

$$\left\{(\phi_i)^T M\,\phi_i\right\}\ddot{a}_i + \left\{(\phi_i)^T K\,\phi_i\right\}a_i = 0 \qquad i=1..n \tag{11}$$

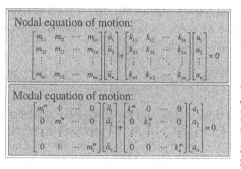

Fig. 15: Modal decomposition

where the terms in the rambled brackets are called modal mass and modal stiffness. Both numbers are computed in a finite element analysis for each mode of interest. The modal decomposition shown here transforms the continuous system which is represented within FEM as an ordinary matrix differential equation system to a series of decoupled single degree of freedom oscillators (Fig. 15). This method is known as modal superposition.

As a matter of fact shape function methods can be extended to nonlinear systems as well [15]. Therefore two approaches are known. First, one performs a modal analysis at several deflection states and obtains deflection dependent but uncoupled mode shapes. Second, one can operate with fixed basis functions (no further eigenvalue extraction is required) but then the modal equation system gets coupled.

The second method is less time consuming and hence applied in the following procedure. A non-linear modal stiffness matrix can be established by the partial derivatives of the strain energy function to the modal coordinate a_i:

$$k_i^m = \frac{1}{a_i} \frac{\partial E_p(a_1, a_2, \cdots, a_n)}{\partial a_i} \tag{12}$$

The accuracy of the final solution depends mainly on the possible precision of the energy function. Typically, the strain energy function of structures with stress stiffening can be expressed by a fourth-order polynomial in the modal coordinates. For a structure with five important modes, it requires about 100 FEM simulations to compute a proper number of data points to fit the coefficients of the energy function. The contribution of each single mode to the total energy function requires a sequence of various combinations of mode shapes which are applied to a finite element model and solved numerically (see Fig. 16). Usually interactions between high-order modes are neglected.

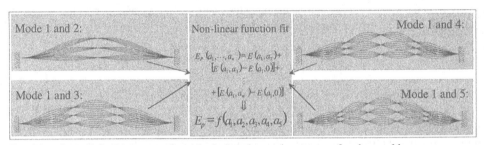

Fig. 16: Sets of displacements for calculating the strain energy of a clamped beam

External mechanical forces or surface pressure can be considered as springs which give the same amount of force at equilibrium ($F = c\,u$). This representation of external loads is convenient since the whole systems can be established by energy equations now. Similar to eq. (10) the potential energy is formulated and the derivative leads to

$$\frac{\partial E_p}{\partial a_i} = \phi_i\, c \sum_{j=1}^{n} \left(a_j\, \phi_j\right) = \phi_i\, c\, u = \phi_i\, F_{ext} \tag{13}$$

where F_{ext} is the external force itself. Since this work is done on the system (i.e. external work) it must be added on the right hand side of eq. (8). As a result of Lagrange's principle each external force has to be multiplied by the i-th shape function amplitude. The physical meaning of this scale factor ϕ_i is that forces acting on the maximum of the eigenmode have the strongest influence to that mode, whereas force close to the vibration node can be neglected. Damping effects are considered in the same way as known from the Rayleigh-approach in numerical mechanics [10].

Including electrostatic forces into our macromodel procedure means that the designer must account electrostatic energy terms to the total potential energy expression. Therefore the mechanical part has to be enhanced by the electrostatic domain similar to Fig. 9. In contrast to the example in Fig. 9 we don't need to perform an iterative solution since the mode shape is completely prescribed. At each deflection state one has to apply a unit voltage to the movable electrode and to summarize the electrostatic field energy.

The generalized force acting on mode i is finally given by

$$\frac{\partial E_{el}}{\partial a_i} = U(t)^2 \frac{\partial E_{el}(a_1, a_2, \cdots, a_n)\big|_{U=1}}{\partial a_i} \tag{14}$$

where $U(t)$ is the real voltage on the electrodes which may be defined later in the macromodel use pass. The general algorithm of macromodel construction can be outlined as follows:

1. Compute the linear modes of the elastic problem,
2. Use nonlinear quasi-static FE-simulations to extract strain energy data over the design space,
3. Develop an analytic strain energy function in terms of selected mode amplitudes,
4. Calculate the generalized mass of each mode and the first derivatives of the strain energy function,
5. Establish the modal equations of motion.

The macromodel generation procedure requires access to a finite element tool and a powerful non-linear function fitting procedure (here the Levenberg-Marquardt method was used). Once the macromodel is generated it can be used for any different load situation either for steady-state or dynamic simulations. The tools which use the macromodel later (use pass) should be able to solve non-linear ordinary differential equations with few unknowns.

The accuracy, robustness and convergence behavior of this macromodeling technique can be proven on a set of examples [16]. The first example is an electrostatically actuated micro-bridge suspended above an electrode strip (see Fig. 17 left). The applied voltage is gradually increased up to pull-in. Since the microstructure is thin compared to the deflection range, growing non-linearities arise. Compared to a finite element solution the macromodel agrees nearly perfectly over the entire voltage range. The transient response of a square plate suspended on four beams is assessed at the right hand side. As known from non-linear systems, the deflection is in general a non-sinusoidal function especially if the driving frequency lies close to high-order modes.

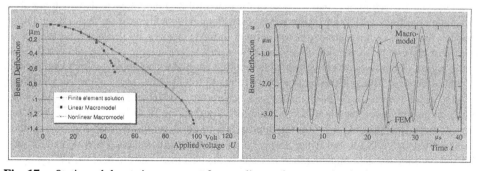

Fig. 17:   Static and dynamic response of a non-linear electromechanical system

An important benefit of modal methods is that they don't require any self-consistent multiple energy domain simulations as referred in chapter 5. The modal basis functions, strain and kinetic energy data can be calculated almost automatically. Numerically pretentious simulation schemes to consider electromechanical interactions can be completely omitted during the generation pass of macromodels. This is possible since the entire deformation state of the mechanical system is prescribed by given displacements and consequently any forces are without effect. By now, the designer has still to decide which modes are relevant and the range of modal amplitudes that should be captured by the macromodel.

There are still limits to the basis function approach. As can be proved, it is difficult to calculate accurate energy terms of an elastic body which undergoes large deflection amplitudes. For instance, trouble occur if the deformed shape of a linear system changes completely at large displacements, such as occurs at contact to the ground electrode. However, the class of devices that can be handled by shape function methods is large enough to be of interest.

7.   Conclusion

In this paper we have presented strategies, tools and algorithms for design and simulation of microelectromechanical components. Measurements on latest commercial microsystems, sensor prototypes and test structures have clearly shown that an accurate and reliable design procedure is possible by now. However, a lot of work remains in the area of

- Material characterization:

Accurate knowledge of material properties is an essential issue for the design. While the behavior of bulk material is well studied, there is a lag on information about thin film properties. It is known that Young's modulus and especially the residual stress of thin films are strongly influenced by film thickness, deposition parameters and subsequent high temperature processing steps. Both material parameters affect the stiffness and natural frequency of the device, multilayer structures bend out of shape and buckling effects may occur. It doesn't make any sense to improve the accuracy of simulation algorithms if on the other hand the material properties are subject to large errors. Modern material parameter extraction techniques require not only experimental efforts but also on-line simulation capabilities [17].

- Efficient 3-dimensional structural modeling:

In microsystems technology, there is a need for powerful 3-D modeling tools which are able to handle structures with high aspect ratios. First demand is addressed on the part modeling module. Volumes of microstructures are defined in CAD systems by surface areas known as boundary representation method. Using bulk micromachining a large number of crystal faces in the region of convex corners is formed by etching simulators but they are irrelevant for the structural behavior. Algorithms and tools are required to clip those areas automatically. Next demand is addressed on finite element mesh generators. For instance, the seismic mass requires in general a completely different mesh density than the flexible suspension. Problems are discussed in chapter 4. A promising approach is to tie dissimilar meshed regions together as supported by ANSYS, but it requires a lot of designer experience. Third demand is addressed on finite element implementation. Usually the designer wants to deal with beam, plate and volume elements in the same model. There are difficulties to link those element types together due to the unlike kind of degrees of freedom. Special interface elements are published in [18].

- Multiple energy domain simulation:

In chapter 5.2 different techniques are presented for simulation of multiple energy domains. Especially in thermomechanics and electromechanics there was enormous progress in the past. In contrast, simulation of fluidmechanical interactions are still a challenge for the designer. The reason is that fluids are described by the Eulerian approach, where the mesh is fixed and the fluid moves. On the other hand solids are considered according the Lagrangian approach where the finite element mesh moves with the deflection. Existing algorithms based on finite element node update are not applicable for fluid flow problems. A new approach using mixed Eulerian-Lagrangian elements is presented in [19] but unfortunately, no commercial simulation tool is able to handle fluid-structure interactions correctly today.

Another problem arise from the iteration scheme which is applied to multiple energy domain simulations. Most FE tools support relaxation techniques by powerful commands but from the mathematical point of view Newton techniques are more efficient and reliable. The trouble with the Newton method is that their solver requires access to the Jacobian matrix of the whole system which is not possible in commercial tools. Interesting alternatives to couple FE tools are presented in [20]. One of it is based on a modified Newton algorithm (called multi-level Newton method) where no system matrix access is required. The non-linear equation system is solved by a number of matrix vector products to minimize the residuum step-by-step (generalized minimal residual algorithm GMRES [21]).

- Robust macromodel extraction techniques:

Macromodel extraction techniques should cover important non-linearities, multiple energy domains and flexible structures. Furthermore a simple to use relationship of input-to-output quantities should be established. Today most macromodels are extracted by numerical techniques (e.g. substructure techniques, shape function methods) since this can be done automated. Unfortunately, there is no access to an analytical relationship of technical design variables such as geometrical dimensions, material properties, applied loads to the output quantities (e.g. deflection, voltage) possible. Numerical parameter extraction techniques which are able to establish mathematical terms of at least the most important design variables will be the goal of future research.

Acknowledgement

This project was sponsored by the DFG (Deutsche Forschungsgemeinschaft) Sonderforschungsbereich 379 and the Federal Ministry for Education and Research (project EKOSAS, V1514).

The authors wish to thank the MEMCAD group of Prof. S. D. Senturia (MIT Boston, USA) for the extensive cooperation in the field of macromodeling and CAD-FEM GmbH Munich for assistance in finite element simulations.

References

[1]   K.D. Müller-Glaser: *Anforderungen der Mikrosystementwicklung*, 4. Workshop of the BMFT project METEOR, VDI/VDE Technologiezentrum Informationstechnik GmbH, Karlsruhe 1996

[2]   R. E: Rudd, J. Q. Broughton: *Coupling of Length Scales and Atomistic Simulation of a MEMS Device*, Boston Area MEMS Seminar, Boston 1998

[3]   J. Mehner, W. Dötzel: *A Design System for Kinetic Micro Mechanical Components*, MICROSIM 95, p. 29-36, Southampton 1995

[4]   S. B. Crary, J. R. Clark, O. S. Juma: *Overview of the CAEMEMS Framework and Optimization of MEMS using Statistical Optimal Design of Experiments*, MICROSIM 95, p. 11-18, Southampton 1995

[5]   J. Wibbeler, C. Steiniger, M Küchler, K. Wolf, T. Frank, J. Mehner, W. Dötzel, T. Gessner: *Resonant Silicon Vibration Sensors with Voltage Controlled Frequency Tuning Capability*, in Proc. EUROSENSORS XIII, 24B3, The Hague 1999

[6]   S. G. Adams, F. M. Bertsch, K. A. Shaw, P. G. Hartwell, F. C. Moon, N. C. MacDonald: *Capacitance based tunable resonators*, J. Micromech. Microeng., Vol. 8,p. 15-23, 1998

[7]   O. C. Zienkiewicz: *Coupled problems and their numerical solution*, John Wiley & Sons Ltd., UK 1984

[8]   L. Yao, E. Chowanietz, M. McCormick: *Design and analysis of a resonant gyroscope suitable for fabrication using the LIGA process*, MICROSIM 95, p. 69-78, Southampton 1995

[9]   L. Weller: *Numerische Mathematik für Ingenieure und Naturwissenschaftler*, Viewegs Verlag 1996

[10] J. Mehner: *Entwurf in der Mirkosystemtechnik*, Dresdner Beiträge zur Sensorik, Band 9, DUB 2000

[11] D. Billep: *Modellierung und Simulation eines mikromechanischen Drehratensensors*, Dissertation, TU Chemnitz 2000

[12] S. D. Senturia: *CAD Challenges for Microsensors, Microactuators and Microsystems*, Proc. of the IEEE, Special Issue on MEMS, June 1998

[13] S. D. Senturia: *CAD for microelectromechanical systems*, J. Micromech. Microeng., Vol. 3, no. 3, p. 103-106, 1993

[14] D. Teegarden, G. Lorenz, R. Neul: *How to model and simulate microgryroscope systems*, Proc. of the IEEE, Special Issue on MEMS, July 1998

[15] K. J. Bathe, S. Gracewski: *On nonlinear dynamic analysis using substructuring and mode superposition*, J. Computers and Structures, Vol.13, p. 699-707, 1981

[16] J. Mehner, L. D. Gabbay, S. D. Senturia: *Computer-Aided Generation of Nonlinear Reduced-Order Dynamic Macromodels: II. Stress-Stiffened Case*, J. Microelectromech. Systems, June 2000 p. 270-280

[17] G. K. Gupta, P. M. Osterberg, S. D. Senturia: *Material property measurement of micromachanical polysilicon beams*, Proc. SPIE 1996, Microlithography and Metrology in Micromachining, Austin, p. 39-45, Texas 1996

[18] K. J. Bathe: *Finite element Methoden*, Springer Verlag 1996

[19] J. Donea, P. Fasoli-Stella and S. Giuliani: *Lagranian and Eulerian finite element for transient fluid-structure interaction problems*, Trans. SmiRT-4, San Francisco 1979

[20] N. Aluru, J.White: *Algorithms for coupled domain MEMS simulation*, Proc. of the design automation conference, Anaheim, USA 1997

[21] Y. Saad, M. H. Schulz: *GMRES: A generalized minimal residual algorithm for solving nonsymmetric linear systems*, SIAM J. Sci. Stat. Comput., Vol. 7, p. 856-869, 1986

A FEM Based Network Approach for the Simulation of Miniaturized Electromagnetic Devices

R. Gollee and G. Gerlach

Dresden University of Technology
Institute for Solid-State Electronics

Abstract

The paper proposes a new computational approach for analyzing miniaturized electromagnetic devices. This approach is based on a general structure of an equivalent network for the coupled electromagnetic problem. It enables automated derivation of the equivalent network by decomposition of the whole device structure into substructures with nearly homogenous field distribution. To assess the actual values of field quantities the device is modeled by 3D-FE simulations. The parameters of the lumped network elements will be obtained by analysis of these results.

1 Introduction

Because of different coupled physical domains (electrical, magnetic, mechanical) the optimal design of electromagnetic devices is rarely possible by direct calculation. Particularly, the decreasing dimensions of electromagnetic devices and the increasing power density in such devices require optimization techniques based on simulation.

So far, simulation based optimization has been limited. Commonly used approaches require extensive and time consuming computational methods such as sequentially coupled physical domains (electromagnetic and mechanical) [1, 2]. Other approaches, using one or more transfer functions, have been limited to invariant geometrical dimensions [3, 4].

Equivalent network approaches have been found very usefull to model whole electromagnetic devices, including variation of their dimensions [5, 6]. The network approach is based on the reduction of distributed vector quantities into scalar quantities by integration over spatial domains with time dependent field quantities which show a constant spatial distribution. The ratios of these scalar quantities represent the parameters of concentrated network elements. To assess the actual values of fields parameters, the device can modeled by 3D-FE simulations. The parameters of the lumped network elements are obtained by subsequent analysis of these FE results.

Approaches to model the mechanical domain with networks routinely have been developed widely [8]. Contrarily, automated generation of models for the electromagnetic domain has not been possible. Up to now, derivation of a suitable structure of the equivalent network has been an empirical process for the electromagnetic domain.

In this regard the following paper introduces the possibility for such an approach for automated derivation and parameter estimation of equivalent electromagnetic networks. It is based on a general network structure for the coupled electromagnetic problem.

In common applications (standard devices such as contactors), only network elements

R. Merker and W. Schwarz (eds.), System Design Automation, 131-139.

representing the iron part show nonlinear behavior, whereas elements representing the stray field will behave linearly. Then, modeling with linear stray field elements usually yields to normally reasonable results. However, for analysis of miniaturized electromagnetic devices with high power density the variation of the stray field's geometry with magnetic feeding has to be considered by using nonlinear stray elements.

The general structure of an equivalent network with nonlinear stray elements, as proposed here, represents a versatile tool for modeling electromagnetic structures. It is valid for all reluctance devices.

2 Derivation of Network Elements

The derivation of lumped network elements from complex 3D electromagnetic devices is based on the decomposition of the whole structure into substructures with simpler geometry (e.g. cuboids, cylinders) and with nearly homogenous field distribution. Vector quantities in a single substructure will be reduced to lumped scalar quantities. In terms of the magnetic field the flux density $\vec{B}$ has to be reduced to the flux ϕ, and the field strength $\vec{H}$ to the magnetic voltage v:

$$\phi = \int \vec{B} d\vec{A}, \tag{1}$$

$$v = \int \vec{H} d\vec{l} \tag{2}$$

($\vec{A}$ – mean cross-section area, $\vec{l}$ – mean length of the substructure). Fig. 1 shows the geometrical interpretation of the reduction. Now, the actual boundaries of the substructure

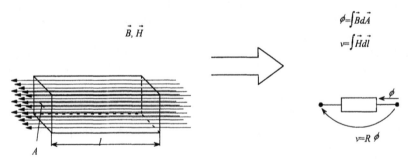

Fig. 1: Geometrical interpretation of the reduction of field quantities to scalar quantities

are reduced to terminals of the network elements except for the magnetic feeding (coil). This feeding will be represented by a magnetic voltage source $\Theta = iw$, where Θ is the magnetomotive force (mmf) and i is the current in the coil with w windings. The network elements can be interpreted as magnetic resistances

$$R = \frac{v}{\phi}. \tag{3}$$

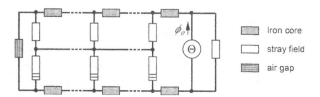

Fig. 2: Structure of the general equivalent network with arbitrary materials assignment

The structure of the network results from the concept of closed magnetic flux lines. Hence, it is possible to define a general network structure which represents any iron core with non branching flux lines (fig. 2). For branching flux lines in the iron domain (e.g. in long stroke linear actuators with more than one air gap) parallel network elements have to be added.

The geometry of the domains representing the iron core and the air gap is known (boundaries of the associated substructures are given by the bounding areas of the iron core and the perpendicular intersections). Therefore, only nonlinear material behavior has to be considered closely for estimation of the parameters of these elements. The resistance of such a network element (index e) can be described by

$$R^e_{Fe}(\mu) = \frac{l^e_{Fe}}{\mu A^e_{Fe}} \qquad (4)$$

using the constitutive law $B = \mu H$ (μ – permeability), eqs. (1) and (2).

The material properties of the elements representing the stray field do not depend on magnetic flux, whereas their geometrical parameters are causing nonlinear behavior of the elements. This fact has been neglected in former analysis of magnetic circuits, e.g. [5] and [6]. The assumption of fully linear behavior of the stray field elements is only valid for analysis of non-miniaturized magnetic devices. For miniaturized magnetic devices, nonlinear behavior of the stray field elements can not be neglected.

The stray field resistance of such an element e can be described by:

$$R^e_s(k^e_s) = \frac{1}{\mu_0 k^e_s} \quad \text{with} \quad k^e_s(H) = \frac{A^e_s(H)}{l^e_s(H)}. \qquad (5)$$

3 Parametrizing of Network Elements

The example of a simple test device (fig. 5) has been chosen to demonstrate the procedure of parametrizing a network derived from the geometry of the iron core. Here, parametrization should be based on the results of 3D finite element analysis (FEA) using ANSYS 5.4. The calculations will be carried out for various mmf Θ (load steps).

In a first step it is necessary to identify and quantify the network elements representing the iron core. For that, the structure of the iron core (main flux region) will be decomposed into substructures Ω_i with nearly homogenous field distribution using an automated procedure as shown in fig. 3. The ratio of incoming lumped flux ϕ^e_1 through area A^e_1 and outgoing lumped flux ϕ^e_2 through area A^e_2 of the structure Ω_i was been proven on satisfying the stop criterion (fig. 4a). If the stop criterion is not satisfied the structure Ω_i must be decomposed into substructures Ω^k_i and Ω^{k+1}_i (fig. 4b). For each substructure

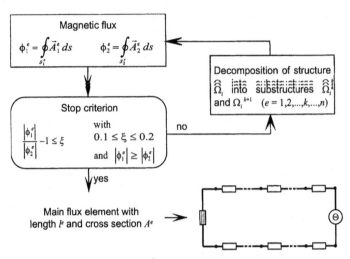

Fig. 3: Automated derivation process of network elements representing the main flux domain

the ratios of incoming and outgoing flux will be compared with the stop criterion. The decomposition of structures ends if the stop criterion is satisfied.

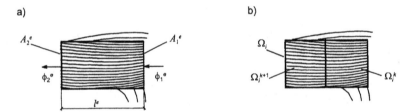

Fig. 4: Flux lines in structure Ω_i; a) Incoming (ϕ_1^e) and outgoing (ϕ_2^e) lumped flux through areas A_1^e resp. A_2^e; b) Decomposition of structure Ω_i into substructures Ω_i^k and Ω_i^{k+1}

Both the geometrical parameter l_{Fe}^e and A_{Fe}^e in eq. (4) as well as the number of network elements result from this procedure. For each load step the flux ϕ^e and the magnetic voltage v^e in the substructures Ω^e are calculated by

$$\phi^e = \oint \vec{a}_e \, d\vec{s}^e \quad \text{with} \quad \vec{a}_e = \text{rot } \vec{B}_e \tag{6}$$

and

$$v^e = \frac{W^e}{\phi^e} \quad \text{with} \quad W^e = \int \vec{B}_e \vec{H}_e \, dV^e \tag{7}$$

($\vec{a}$ – magnetic vector potential, W^e – magnetic energy, V^e – volume).

The subnet for the iron core enables the subsequent specification of the stray field elements. Considering the equivalent network in fig. 2 the fluxes ϕ_s^e and magnetic voltages v_s^e of the stray field elements are obtained from applying Kirchhoff's laws. The actual geometrical parameters l_s^e and A_s^e in eq. (5) can not assessed due to the complexity and varying dimensions (as stated above) of the stray field. Nevertheless, Kirchhoff's laws and the energy conservation are ensured using a parameter k_s^e, depending on magnetic field strength. It can be calculated by eqs. (3) and (5).

For numerical handling a continous function describing the change of the parameter k_s^e has to be found. For this, element specific parameters k_0^e will be extracted from FEA results and k_s^e can be written as

$$k_s^e(H) = k_s(H)k_0^e, \tag{8}$$

where $k_s(H)$ is a general continous function which describes the nonlinear behavior of the stray field elements. It can be approximated by spline interpolations.

For the numerical solution of the system of nonlinear equations a Newton-Raphson algorithm has been used (see section 4). Therefore, a potential function

$$b(H) = \int k_s(H)dH \tag{9}$$

was necessary for the consistent description of the nonlinearity.

4 Numerical Solution

The solution of the parametrized equivalent network is based on the principle of minimizing the total magnetic energy [7]:

$$\sum_{e=1}^{n} W^e \to \min \tag{10}$$

(n is the number of network elements).

For one-dimensional network elements (index e) with two terminals a linear approach can be used to describe the magnetic voltage v depending on its length l within the elements:

$$v^e(l) = a_1 + a_2 l = \mathbf{u}_e^T \mathbf{a}$$
$$\text{with } \mathbf{u}_e = \begin{Bmatrix} 1 \\ l \end{Bmatrix} \text{ and } \mathbf{a} = \begin{Bmatrix} a_1 \\ a_2 \end{Bmatrix}. \tag{11}$$

The magnetic voltage potentials at the terminals will be described by a column matrix $\mathbf{v}_e$:

$$\mathbf{v}_e = \mathbf{A}_e \mathbf{a} = \begin{bmatrix} 1 & l_1 \\ 1 & l_2 \end{bmatrix} \begin{Bmatrix} a_1 \\ a_2 \end{Bmatrix} \tag{12}$$

with $l_1 = 0$ and $l_2 = l^e$. Now, the vector of coefficients $\mathbf{a}$ will be eliminated using eq. (12), and eq. (11) can be written as

$$v^e = \mathbf{u}_e^T \mathbf{a} = \mathbf{u}_e^T \mathbf{A}_e^{-1} \mathbf{v}_e. \tag{13}$$

Using a differential operator $\Gamma = \frac{d}{dl}$ and eq. (2) the field strength H^e is

$$H^e = \Gamma v^e = \Gamma \mathbf{u}_e^T \mathbf{A}_e^{-1} \mathbf{v}_e. \tag{14}$$

The magnetic energy inside the element e given by the magnetic field strength H^e (inner energy) results from

$$
\begin{aligned}
W_i^e &= \int_0^{l^e} (H^e)^2 \, E^e \, dl \\
&= \mathbf{v}_e^T \, \mathbf{A}_e^{-1\,T} \underbrace{\int_0^{l^e} \mathbf{c}_e^T E^e \, \mathbf{c}_e \, dl \, \mathbf{A}_e^{-1}}_{\mathbf{G}_e} \mathbf{v}_e
\end{aligned}
\tag{15}
$$

with $\mathbf{c}_e = \Gamma \mathbf{u}_e^T$ and $E^e = \mu^e A^e$. $\mathbf{G}_e$ represents the magnetic conductivity matrix. The magnetic energy given by the imposed magnetic flux ϕ^e (outer energy) is directed opposite the inner energy flux:

$$
W_o^e = -\int_0^{l^e} q^e \, v^e \, dl = -\int_0^{l^e} q^e \mathbf{u}_e^T \mathbf{A}_e^{-1} \mathbf{v}_e \, dl = \boldsymbol{\phi}_e^T \mathbf{v}_e,
\tag{16}
$$

where $q^e = d\phi^e/dl$ and $\boldsymbol{\phi}_e$ is the implied flux. The total magnetic energy in the network element e is the sum of the inner and the outer energy:

$$
W^e = W_i^e + W_o^e = \mathbf{v}_e^T \mathbf{G}_e \mathbf{v}_e - \mathbf{v}_e^T \boldsymbol{\phi}_e.
\tag{17}
$$

The assignment of element values to global values will be done using incidence matrices $\mathbf{I}_e$ of the type $(2, k)$ with $k = n/2 + 2$ as the number of network nodes. With

$$
\mathbf{v}_e = \mathbf{I}_e \mathbf{v}
\tag{18}
$$

the total energy of the system can be written as

$$
\begin{aligned}
W &= \mathbf{v}^T \left(\sum_{e=1}^n \mathbf{I}_e^T \mathbf{G}_e \mathbf{I}_e \right) \mathbf{v} - \mathbf{v}^T \sum_{e=1}^n \mathbf{I}_e^T \boldsymbol{\phi}_e \\
&= \mathbf{v}^T \mathbf{G} \mathbf{v} - \mathbf{v}^T \boldsymbol{\phi}.
\end{aligned}
\tag{19}
$$

The minimum of the total energy results from

$$
\begin{aligned}
\frac{dW}{d\mathbf{v}} &= \frac{d \left(\mathbf{v}^T \mathbf{G} \mathbf{v} - \mathbf{v}^T \boldsymbol{\phi} \right)}{d\mathbf{v}} \\
&= \mathbf{G}\mathbf{v} - \boldsymbol{\phi} = 0.
\end{aligned}
\tag{20}
$$

Hence, the system of equations become

$$
\mathbf{G}\mathbf{v} = \boldsymbol{\phi}.
\tag{21}
$$

Because of the nonlinear behavior of both types of network elements, the solution of this system of equations has to be obtained with a nonlinear solver. For this, a modified Newton-Raphson algorithm was implemented, in which the applied load (mmf Θ) will be increased stepwise up to the desired value. At all steps full Newton-Raphson iterations will be performed for a convergent solution.

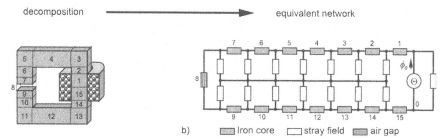

Fig. 5: Analyzed test device: a) decomposed iron core; b) resulting equivalent network

5   Results

The new approach has been demonstrated for the device shown in Fig. 5a). The iron domain was decomposited into fifteen parts. Resulting from the general structure in fig. 2, a network of thirty lumped elements has been obtained (Fig. 5b).

To parametrize the lumped network elements several 3D Finite Element Analysis were carried out, varying the mmf Θ. The behavior of the iron representing elements and the stray field elements are shown in fig. 6.

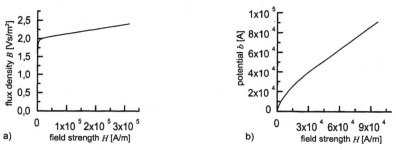

Fig. 6: Behavior of the network elements: a) iron core; b) stray field

The magnetic conductivity matrices

$$\mathbf{G}_e = \begin{bmatrix} \frac{1}{R^e} & -\frac{1}{R^e} \\ -\frac{1}{R^e} & \frac{1}{R^e} \end{bmatrix} \quad \text{for } (e = 1 \ldots 30) \tag{22}$$

and incidence matrices $\mathbf{I}_e$ of the type $(2, 17)$ yield the system of equations (21). The global magnetic voltages $\mathbf{v}$ regard to point 0 in fig. 5.

Fig. 7 shows the comparison of fluxes ϕ in selected elements obtained by the approach described here and by measurement for various applied mmf Θ. The correspondence between these data is quite satisfactory.

6   Conclusion

The described method enables automated derivation and parameter estimation of electro-magnetic networks. Based on FEM results complex magnetic subsystems can be described

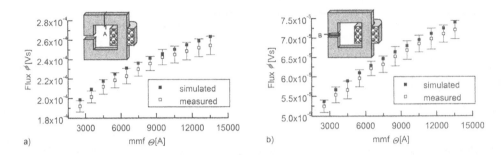

Fig. 7: Simulated and measured flux ϕ vs. mmf Θ: a) inside the iron core A, b) in the air gap B

by a general equivalent network. The nonlinear behavior of the elements both representing the iron core and that of the stray field elements has been included. The resulting system of nonlinear equations can be solved using a Newton-Raphson algorithm.

The basic geometrical dimensions of the iron core are parameters of the network elements. This enables geometrical variations of the electromagnetic device.

Calculating electromagnetic forces (e.g. by the approach of virtual displacement), the electromagnetic network can be coupled to a mechanical equivalent network. By that, the dynamic response characteristic of an electromagnetic actuator can be described. Miniaturized electromagnetic devices can be analyzed by the presented new method especially.

References

[1] Sadowski, N.; Bastos, J. P. A.; Albuquerque, A. B.; Pinho, A. C. and Kuo-Peng, P.: A voltage fed AC contactor modeling using 3D edge elements. IEEE Trans. on Mag., Vol. 34, No. 5, Sept. 1998, pp. 3170-3173

[2] Srairi, K. and Féliachi, M.: Numerical coupling models for analyzing dynamic behaviors of electromagnetic actuators. IEEE Trans. on Mag., Vol. 34, No. 5, Sept. 1998, pp. 3608-3611

[3] Gollee, R.; Roschke, Th. and Gerlach, G.: A finite element method based dynamic analysis of a long-stroke linear actuator. J. of Magnetism and Magnetic Mat., Vol. 196-197, May 1999, pp. 943-945

[4] Sadowski, N.; Carlson, R.; Beckert, A. M. and Bastos, J. P. A.: Dynamic modeling of a new designed linear actuator using 3D edge element analysis. IEEE Trans. on Mag., Vol. 32, No. 3, May 1996, pp. 1633-1636

[5] McDermott, Th. E.; Zhou, P.; Gilmore, J. and Cendes, Z.: Electromechanical system simulation with models generated from finite element solutions. IEEE Trans. on Mag., Vol. 33, No. 2, March 1997, pp. 1682-1685

[6] Roschke, Th.; Gerlach, G.: An equivalent network model of a controlled solenoid. IASTED Internat. Conf. on Applied Modelling and Simulation, Banff, Canada, July 27-Aug. 1, 1997, pp. 241

[7] Engeln-Hüllges, G.; Reutter, F.: Numerik-Algorithmen. 8th Ed., VDI-Verlag, Düsseldorf, 1996

[8] Lenk, A.: Elektromechanische Systeme. Verlag Technik, Berlin, 1975.

High Performance Control System Processor

René A. Cumplido-Parra[1], Simon R. Jones[1], Roger M. Goodall[1] and Stephen Bateman[2]
[1]Loughborough University, UK; [2]Gatefield Corporation, USA

Abstract

This paper describes a compact, high-speed special purpose processor, which offers a low-cost solution to implement linear time invariant controllers. The controller has been reformulated into a modified state-space representation based on the δ operator, which is optimised for numerical efficiency. This Control System Processor (CSP) has been implemented using a programmable ASIC (ProASIC) device.

1 Introduction

There is now a variety of control design methods by which appropriate control laws can be created for complex multi-variable systems, but the actual implementation of control laws is a part of the design process which most control engineers want to achieve as straightforwardly and transparently as possible. One approach is to programme a fixed point microprocessor (μP) device in a high level language, using floating point variables so that numerical issues are not a concern, but the computational overhead is large and surprising restrictions in sample rate are found. A second approach is to use a fixed point digital signal processor (DSP) which has an architecture better targeted for computationally-intensive applications, and these offer some speed advantage over a μP. Other options are to use a number of parallel processors or sophisticated floating-point DSPs, but this doesn't result in a cost-effective solution, especially for high-volume embedded control applications.

The difficulty is that there are particular numerical requirements in control system processing for which standard processor devices are not well suited, in particular arising from the high sample rates which are need to avoid adverse effects of sample delays upon stability. These could be satisfied in either μP or DSP devices using "hand-crafted" numerical routines, probably written in assembler language, but as mentioned above control engineers generally have neither the will nor the skill to do this. There is therefore a clear need to understand the numerical requirements properly, to identity optimised forms for implementing control laws, and to translate these into efficient processor architectures.

Continuing improvements in microelectronic technology has made feasible large reprogrammable silicon chips (Field Programmable Gate Arrays – FPGAs) which can be configured to realise complex computational systems without incurring either the delay or the costs associated with custom silicon. As a result electronic designers and control engineers are looking once again at the potential of designing low-cost, high-performance special-purpose hardware for embedded real-time control. With current chip complexities of up to 2 million designable gates and circuit density growth of 100% every 18 months it appears that the complexity of algorithm is limited only by design capability and not by silicon complexity.

140

R. Merker and W. Schwarz (eds.), System Design Automation, 140-151.
© 2001 *Kluwer Academic Publishers.*

1.1 The use of targeted architectures

It is well accepted that customisation of silicon offers cost and performance advantages over standard components. Furthermore FPGAs open up this market to a much smaller product volume. Contemporary microprocessors offer high-performance at the price of increased cost and power consumption. Furthermore, there is good evidence that the extensive use of floating-point numbers in calculation results in large chips and slow operation. Our view is that be taking a considered view of the numerical and calculation requirements of the algorithm allows special purpose processors to be considered which provide well-targeted support of control laws. Such systems are likely to be smaller, cheaper, faster and lower power than conventional signal processors. In the past there has been much work on special purpose processors for control (e.g. Jaswa's CPAC [1], PACE [2]) but while they are intriguing ideas the cost of producing custom silicon proved prohibitive for initial exploitation and restricted experimentation with different architectural constructs. With High-level design tools such as VHDL and logic synthesis CAD suites allied to large low-cost reprogrammable FPGAs, the constraints no longer apply and we can now develop this area with full enthusiasm.

The remainder of this paper is structured as follows. Section 2 reviews the state-space representations of a controller, describes the formulation used to implement the CSP and the associated numerical issues. Section 3 describes the CSP architecture and its core. Section 4 details the CSP instruction set, program structure and software suite. Section 5 presents some simulation results, and finally section 6 concludes.

2 Control Issues

2.1 State-Space equations

Two approaches are available for the analysis and design of feedback control systems. The first is known as the classical, or frequency-domain, technique. This approach is based on converting a system's differential equation to a transfer function, thus generating a mathematical model that algebraically relates a representation of the output to a representation of the input. The primary disadvantage of the classical approach is its limited applicability: it can be applied only to linear, time-invariant systems or systems that can be approximated as such.

With the arrival of modern applications, requirements for control systems increased in scope. Modelling systems by using linear, time-invariant differential equations and subsequent transfer functions became inadequate. The state-space approach is a unified method for modelling, analysing, and designing a wide range of systems.

Although this representation of the system still involves a relationship between the input and output signals, it also involves an additional set of variables, called state variables. The mathematical equations describing the system, its input, and its outputs are usually divided in two parts: a set of mathematical equations relating the state variables to the input signal and a second set of mathematical equations relating the state variables and the current input to the output signal.

The state variables provide information about all the internal signals in the system. As a result, the state-space description provides a more detailed description of the system than the input-output description. Many systems do not just have a single input and a single output. Multiple-input, multiple output systems can be compactly represented in state space with a model similar in form and complexity to that used for single-input, single-output systems. The general form of the state-space equations, to which all forms of control systems can be converted, is:

$$X_{N+1} = AX_N + BU_N \qquad (1)$$

$$Y_N = CX_N + DU_N \qquad (2)$$

2.2 Formulation used for the CSP

To implement the control algorithm we decided to adopt a modified structure based on the delta operator which present a number of advantages; among them, it does not require the long coefficient word-lengths needed to cope with high coefficient sensitivity associated with the z-operator.

The numerical problems associated with discrete-time control, in which the sample frequency will typically be two orders of magnitude higher than the controller bandwidth, are well known when the z operator is used [3]. This problem arises specially when sampling at high speed is needed, mostly because the difference between current and next input and output values may be increasingly small. Similarly it is recognised that the use of the δ operator overcomes a number of these problems, in which case the state equation becomes

$$\delta X = A_\delta X + B_\delta U \qquad (3)$$

This is sometimes defined as $\delta = (z-1)T$ (where T is the sample period), in which case there is a unification between discrete and continuous time since $\delta \to s$ (the Laplace operator) as $T \to 0$ [4]. In fact for the relatively high sample frequencies found for practical controllers $\delta = s$ is quite a realistic approximation and the coefficients in A_δ and B_δ become almost independent of the sample period. The effect of sample period must of course be taken into account when implementing δ, and an alternative simpler definition is to use $\delta = z-1$ [5], in which case the correspondence between δ and s is lost but implementation is more direct

The modified canonic δ form affects the representation of the A and B matrices used to calculate the next value of the state variables. A large number of zeroes present in the A matrix can reduce the overall computation time, this is achieved by expanding and rearranging the matrices to reduce the number of operation required. The general form of the actual control equations, which will be implemented, is

$$\begin{aligned} X(n+1)T &= AX(nT) + BU(nT) \\ Y(nT) &= CX(nT) + DU(nT) \end{aligned} \qquad (4)$$

The modified canonic δ form is illustrated diagrammatically in figure 1 for a fourth-order SISO controller.

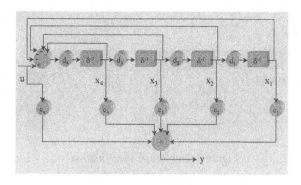

Figure 1. Modified canonic δ formulation

The corresponding state equations are:

$$\begin{bmatrix} x_1 \\ x_2 \\ x_3 \\ x_4 \end{bmatrix}_{n+1} = \begin{bmatrix} 1 & d_1 & 0 & 0 \\ 0 & 1 & d_2 & 0 \\ 0 & 0 & 1 & d_3 \\ 1-d_4 & -d_4 & -d_4 & -d_4 \end{bmatrix} \begin{bmatrix} x_1 \\ x_2 \\ x_3 \\ x_4 \end{bmatrix}_n + \begin{bmatrix} 0 \\ 0 \\ 0 \\ d_4 \end{bmatrix} u_n \qquad (5)$$

$$y_n = \begin{bmatrix} c_1 & c_2 & c_3 & c_4 \end{bmatrix} \begin{bmatrix} x_1 \\ x_2 \\ x_4 \\ x_4 \end{bmatrix}_n + \begin{bmatrix} c_u \end{bmatrix} u_n$$

and the actual equations used for real-time implementation are as below; firstly the calculation of the output, then an update of the states ready for the next sample:

$$y := c_1 x_1 + c_2 x_2 + c_3 x_3 + c_4 x_4 + c_u u$$
$$x_{temp} := x_1 + x_2 + x_3 + x_4$$
$$x_1 := x_1 + d_1 x_2$$
$$x_2 := x_2 + d_2 x_3 \qquad (6)$$
$$x_3 := x_3 + d_3 x_4$$
$$x_4 := x_4 - d_4 x_{temp} + d_4 u$$

Notice that x_{temp} is used to store the sum of the old values of x_1 to x_4, which thereby avoids having to retain old values for the states while the new values are calculated – the state variables are then simply overwritten at each calculation.

2.3 Numerical requirements

Using this standard controller formulation the requirements for coefficients and controller state variables become relatively standardised across a wide range of applications, and these are illustrated in figure 2.

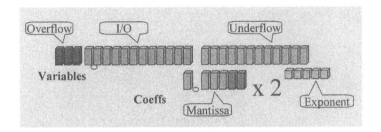

Figure 2. Numerical specification

The variables are 27 bit fixed point, with the input values brought in as integers, a small 3-bit allowance for overflow (although this is a nominal requirement because of the good scaling properties mentioned previously), and 12 fractional bits for underflow.

The coefficients are held in a simple low-precision floating-point form, with 6 bits for the mantissa and 5 bits for the exponent. In general the coefficients fractional with values which become progressively smaller as the sample frequency is increased, but a positive exponent is provided to implement greater than unity gains, a few of which are associated with most controllers.

This numerical specification will implement successfully the vast majority of LTI controller examples, and allows for the sample frequency being at least three orders of magnitude higher than the lowest pole in the controller. Of course if there are exceptional requirements it is always possible to reprogram the CSP hardware in the FPGA, maintaining the essential principles but extending the hardware precision as required. An example of implementing extremely high sample frequency digital filters using the modified canonic δ approach can be found in [6].

3 Hardware Architecture

3.1 CSP architecture

Figure 3 shows a block diagram for the CSP system. The core of the CSP comprises a simplified datapath with storage and computation capabilities. The Register bank stores all the constants, coefficients, state variables, inputs data, output data and partial products needed to perform the calculations.

The computation of the output values is done iteratively executing multiply-accumulation (MAC) operations. These operations are specified by the instructions fetched from an internal program ROM and decoded by the instruction handler. The instruction format contains the source and destination addresses of the operands used in the MAC operation. The program counter addresses the next instruction in program memory to be executed. An internal Data ROM contains the coefficients and the initial values for the state variables and program counter registers. A group of analogue-digital and digital-analogue converters provide the interface to the physical system being controlled. Figure 4 shows the processor interface, grey lines indicate data values while black lines indicate control signals.

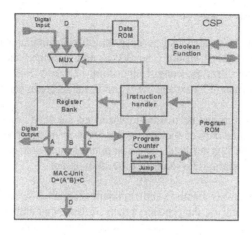

Figure 3. Processor architecture

The processor will be embedded within the complete control system and will normally be programmed in a separate programming system.

Important features of this architecture are:
- reduced precision of the variables when compared to full IEEE floating-point representation

- different numerical representations of coefficients and state variables which are satisfactory for a wide-range of controllers

- targeted MAC unit optimised to for calculating sum of products

This novel architecture combined with the use of a small and specialised instruction set presents cost and performance benefits for control applications over traditional architectures.

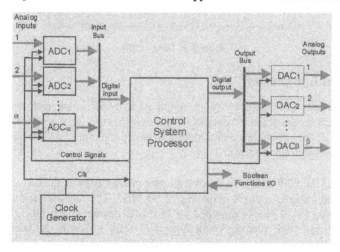

Figure 4. Processor interface

3.2 Core description

The core of the CSP includes a special-purpose multiply-accumulator unit (MAC) and a 4-port register bank (3 read, 1 write). The MAC unit executes the multiply-accumulate operations required to perform the control algorithm, i.e. D=A*B+C (see figure 5). The A input is in coefficient format (12 bits) and the B and C values are in variable format (27 bits).

A detailed low level design has been used to speed up the MAC operation. The system is pipelined such there is a latency of 4 clock cycles between instructions issues and the result being written back to the register bank. The compiler ensures that instruction dependencies are observed through an appropriate series of instruction issues.

The coefficient is split into its mantissa and exponent sections. The multiplier block multiply a state variable by the mantissa, the product is then shifted by a number of bits determined by the coefficient exponent. Finally, the result is added to other state variable to produce the output.

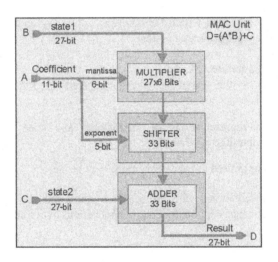

Figure 5. MAC Unit

3.3 Speed and Complexity

Table 1 shows the CSP complexity in terms of ProASIC tiles and equivalent gates. Everything except the program and data ROM are fixed in size; these are hardwired, and their size and speed depends upon the control algorithm being implemented. The figures shown are for the controller specified in section 2.2.

The synthesis of the CSP results in an overall gate count of fewer than 21,000 gates and a delay of 20ns, this allows a clock frequency of 50 MHz. The register bank is implemented using 9 embedded RAM blocks provided by ProASIC devices. Each block contains a 256 word deep by 9 bits wide memory, with 2 ports (1 read, 1 write). As such the CSP is a compact low cost core capable of implementing the most demanding real-time systems.

Block	Tiles (ProAsic)	Equivalent gates
Instruction Handler	101	808
MAC Unit	1105	8840
Program counter	175	1400
I/O Block	60	480
Pipeline registers	120	960
Program ROM	900	7200
Data ROM	80	640
Total	2541	20328

Table 1. CSP complexity

The maximum sampling frequency for a specific control system is determined by its complexity, i.e. the number of instructions needed to calculate the next state and output values.

The relatively small size of the processor core leaves much of the FPGA free such that it can be used to carry out other functions typically associated with real-time control – logical interlocking functions, background tasks such as gain-scheduling etc.

4 CSP Software

4.1 Instruction Set

Now we describe the operation of the CSP from the programmer's point of view. The CSP instruction set is very simple and specialised; it is targeted towards high-speed computation. Due to the MAC unit contains one pipe stage, multiple instructions can be overlapped in execution. At the time that the operands specified by one instruction are being read from the register bank and copied to the MAC unit inputs, the results obtained from the previous instruction is obtained at the output of the MAC unit and copied back to the register bank.

The processor only has four instructions (see table 2). The MAC instruction executes a multiply-accumulation operation on the operands indicated by the source addresses and stores the result in the destination address. This instruction allows performing the matrix multiplication accordingly to the state-space representation of the control system. Because the constants 0, 1 and -1 are stored in the register bank, other calculations can be mapped into this format. The MAC instruction can be used to add two values, increment a value by one, invert the sign of a value, simulate a no operation instruction, copy the contain of one regis7ter, etc. (e.g. D=C is achieved by setting A to 0).

There are no conditional jumps in the system. Unconditional jumps are supported for user programming. However, the code generator flattens all but the exterior loop. The program counter starts at zero increments until it reaches the value stored in the 'jump1 register', it is then reset to the value in the 'jump2 register'. This permits an initial start up sequence to be followed and then a main loop to be repeatedly executed.

The READ instruction allows loading initial values from the data ROM into the register bank during the initialisation process. Also, when the algorithm loop has begun, this instruction is used to read sampled input data from the input bus and to indicate the conversion time for the ADC's. Finally, the WRITE instruction is used to transfer the output values to the output bus.

The program is stored in the same chip as the processor together with simple Boolean functions for interlocking etc. A/D and D/A interfaces are also provided.

Instruction	Function	Description
MAC d, s1, s2, s3	(s1*s2) + s3 -> d	Multiply accumulation operation
WRITEPC sel, s	s -> pc[sel]	Write to program counter registers (PC,jump1,jump2)
READ d, in_pt, s	If in_pt =0 DataRom[s] -> d else Input[in_pt] -> d	Read from Data ROM or input ADC
WRITE out_pt, s	s -> output[out_pt]	Read from Register Bank and write to output DAC

Table 2. CSP instructions

4.2 Program Structure

To generate the CSP Program it is necessary consider the order in which operations must be done, number of inputs, number of outputs and order of the control system to be implemented. Although the CSP program is modified accordingly to the system to be controlled, the program scheme remains the same. It is divided in two main parts: initialisation and algorithm loop. In the initialisation part, all the coefficients and state variable initial values are transferred from the external data ROM to the register bank. Also, the program counter registers that specify the initial and final instruction for the algorithm loop are updated. Finally, the program enters an infinite loop where the input samples are used to calculate the output values to the control system. One of the main goals of this software scheme is to achieve a constant sample rate. The sub-tasks contained within the CSP program are listed below. Additionally, the numbers of CSP instructions required to perform each task are provided.

Task	Number of CSP instructions
Load Coefficients	$n + n\alpha + \beta n + \alpha\beta + 3$
Initialise State Variables and PC	$n + 2\beta + \alpha + 6$
: *Algorithm cycle start*	
Get Input Data	$\alpha + 2$
Calculate DU_N	$\alpha\beta$
Calculate Yn	β
Calculate X_N	$(2 + \alpha)n$
Calculate AX_N	
Calculate BU_N	
Calculate CX_{N+1}	$n\beta$
Supplementary Processes	$\beta + 1$
Write Output Data	
: *End Algorithm cycle*	

where α, β and n are defined as the number of input channels, the number of output channels and the number of internal state variables respectively.

4.3 Software suite

4.3.1 CSP model

The purpose of the CSP Model is to provide a clear understanding of the algorithm and its numerical requirements, as well as a verified functional specification of the processor and test vectors to verify the hardware design. The CSP model architecture is modular; this modularity allows us to replace processing elements to perform performance comparisons and to explore new architectures. Furthermore, the model supports alternative algorithms thus making it suitable for demonstration purposes in a range of control application environments. Input data to the model is provided from Matlab simulations or from the CSP signal generator and the program to be simulated is generated by the CSP program generator.

4.3.2 CSP signal generator

One of the basic analysis and design requirements is to evaluate the response of a system for a given input. Test input signals are used, both analytically and during testing, to verify the design of a control system. It is not practical to choose complicated input signals to analyse performance. Thus, usually standard test inputs are used. These inputs are impulses, steps, ramps, parabolas and sinusoids. The CSP signal generator provides input test data to the CSP model. The signal streams are stored in binary files using double format. The type of signal, number of samples, magnitude, sample frequency, etc. are indicated by parameters sent to the program in the command line. The double format used to store data in the files is compatible with Matlab double formats, so these test files can be used as inputs in any part of the design process.

4.3.3 CSP program generator

The CSP program is created using the program generator. The sequence of instructions needed to perform the control algorithm is that illustrated in section 4.2. The number of instructions varies accordingly with the number of inputs, outputs and order of the system. Each calculation part is generated using these parameters to modify the source and destination addresses for each instruction. Also, the program generator rearrange the order in which the instructions are executed to avoid the data dependency problems associated with pipelined designs.

4.3.4 Programming the CSP

When the CSP program is correct, code can be generated in text format. This code is then converted into VHDL (as a ROM element) and added to the VHDL code. This is then synthesised and placed and routed. Note that this implies that system clock speed can be influenced by program complexity. The chip is then placed in the system board and operation commences on an asserted start signal. The processor is implemented in an Actel ProAsic FPGA (figure 6) which is a flash-programmable non-volatile device. Figure 7 shows the programmer and test system.

Figure 6. Actel ProASIC device   Figure 7. Programmer and test system

5 Results

Results of some simulations of the CSP are shown. These results have been cross-referenced against the results obtained with a Matlab program. Consider a fourth order 1Hz Butterworth filter as that of section 2.2 with a sample frequency of 100Hz. The coefficient values are:

$$d_1 = 0.022, \quad d_2 = 0.0446, \quad d_3 = 0.088, \quad d_4 = 0.178$$
$$c_1 = c_2 = c_3 = c_u = 0 \quad \text{and} \quad c_4 = 1$$

Two normalised input data stream were used in the simulations (i.e. $VE_{min} = -1$ and $VE_{max} = 1$).

1. Step input $u(t) = \begin{cases} 0.5 & t > 0 \\ 0 & \text{otherwise} \end{cases}$

2. Sin input $u(t) = \begin{cases} 0.5 * \sin(t) & t > 0 \\ 0 & \text{otherwise} \end{cases}$

Figures 8 and 9 show the output and state variable values of the CSP VHDL Model and Matlab program with the step and sine signal as input respectively. Clearly, the values obtained with the two approaches are very similar. This demonstrates that the CSP specification is correct and can be implemented in practical applications.

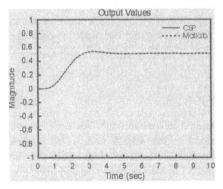

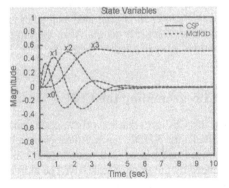

Figure 8. Output values and state variables for input signal 1

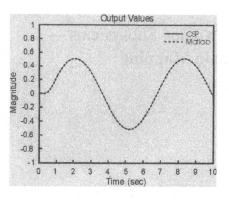
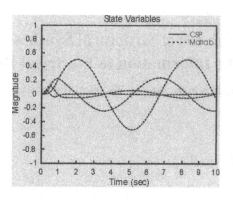

Figure 9. Output values and state variables for input signal 2

The time to perform a single MAC operation is one clock cycle, which is 20ns. The number of instructions needed to perform an algorithm cycle, which includes calculating the next state variables and outputs values is 23 (see section 4.2). Thus the total time required per algorithm cycle is 0.46 μs. This gives a maximum sampling frequency of 2.16MHz.

6 Conclusions

In this paper has been shown a special-purpose control system processor as a solution to implement complex linear time invariant controllers. The processor is implemented in an Actel ProAsic FPGA which is a flash-programmable device (non-volatile) together with the appropriate programmer, offering a one-chip solution. Further work is under way to construct a practical CSP demonstrator and test the design in a range of applications, including MAGLEV suspension control and aircraft flight control.

7 References

[1]   Jaswa, V.C., Thomas, C.E., & Pedicone, J. (1985). CPAC: Concurrent processor architecture for control. IEEE Transactions on computers, C-34, 163-169.

[2]   Spray, A., & Jones, S. (1991). PACE: A regular array for implementing regularly and irregularly structured algorithms. IEE Proceedings, Pt G, Vol. 138, No. 5, pp 613-619.

[3]   Liu B, "Effect of finite wordlength on the accuracy of digital filters – a review", IEEE Trans Circuit Theory, 1971, CT-18, (6), pp 670-677.

[4]   Middleton R H and Goodwin G C, "Digital control and estimation – a unified approach" (Prentice Hall, 1990)

[5]   Goodall R M and Brown D S, "High speed digital controllers using an 8-bit microprocessor", Software and Microsystems, Vol. 4, Nos. 5 & 6, pp 109-116, Dec 1985.

[6]   Goodall R M and Donaghue B, "Very high sample rate digital filters using the operator", Proceedings IEE, Pt G, Vol. 140, No 3, pp 199-206, June 1993.

High Level Structural Synthesis of Analog Subsystems — The Information to Electrical Domain Mapping

Jürgen Kampe, Ilmenau Technical University, Germany*

Abstract

High level design representations are oriented towards information processing, and enables structural synthesis steps on the basis of directed information flow graphs. From the electrical energy flow point of view, electrical circuits are synthesized by using electrical engineering approaches. This paper presents a methodology to over-step the gap between high-level, and electrical design representations. The transformation of the directed signal parameter flow into the electrical domain is performed by a formal approach. The description of an analog implementation of a changeover switch serves as example for the specific design flow.

1 Introduction

The interfaces of electronic information processing systems are mostly composed of analog modules for performing data and signal conditioning.

Early work in the late 1980's, to structurally synthesize analog blocks was based on architectural selection [1][2]. These tools were inflexible to deal with a large variety of specifications. In connection with hierarchical decomposition, design automation tools became more flexible in the early 1990's [3]–[12]. These tools were restricted to specific classes of analog circuits such as opamps and data converters. Further, it was difficult to introduce new circuit structures. Besides simulation based methods [13][12][14], analytical [2][15] and formal [16][17] approaches were established for sizing, specification refinement, and the evaluation of specific topologies. Next, formal techniques were introduced [18] to the process of structural synthesis. Formalization leads to more flexible synthesis and overcomes the restriction to rigid circuit classes by focusing on signal processing and abstract design representations.

Formal techniques enable design approaches starting from algorithmic, or data flow specifications [19]. From the abstraction point of view, data flow design representations give the basis to both digital signal processing and analog circuit implementations and enables high-level design approaches to the structural synthesis of analog circuits. Thus, the focus is given to data processing analog components, such as sensoric front ends, and to communication systems.

In [20], a design strategy is introduced, which is based on a multi-phase design flow, separating the design of the data flow and the design of the electrical signal flow. All steps required for

* Supported by the "Deutsche Forschungsgemeinschaft SFB 358"

R. Merker and W. Schwarz (eds.), System Design Automation, 152-163.
© 2001 *Kluwer Academic Publishers.*

the design of the data flow belong to the analog functional synthesis and all steps necessary for the design of the energy flow are part of the electrical synthesis. In this design methodology, a mapping of the high-level design representation to the electrical domain is required.

Here, the implementation of the high-level to electrical domain mapping process is introduced. The mapping process is based on a appropriate modeling in the electrical domain, giving the design primitive knowledge base; and it is based on a formal description of the design steps, giving the synthesis step knowledge base.

2 The Design Flow

A multi-phase design flow enables us to split both the analog functional synthesis process and the electrical synthesis process into two phases: A functional driven design phase and a sizing oriented design phase.

In the functional driven design phases structural selection and decomposition is done on the basis of the systems or circuits function and on the basis of classified properties of the data or the signals, corresponding to the level of abstraction. Thereby, only main principles are to be taken into account; more complex structures are established by means of formal methods.

The sizing-oriented design phases are starting from symbolic relations, expressing functional relations between the elements. According to the level of abstraction, data or signal parameter values are considered, derived from the design specifications. Additionally, sizing-dependent structural transformations are applied. On the one hand, these transformations take into account the dependency of structural selection from the specific sizing and, on the other hand, these transformations drastically reduce the number of alternatives that has to be taken into consideration in each structural selection step.

Fig. 1 gives an overview on the design steps.

3 Design Representations

3.1 Formal Design Representations

Formal representations of structural and behavioral design primitives are introduced here, defining classes of design primitives at a given level of abstraction.

In general, a behavioral design representation is specified by a set of *n terms* $H_1, ..., H_n$, giving the white box description of the design element. From the structural point of view, a design is represented by a set of *k aspects* $r_1, ..., r_k$, defining structural elements within the design and characterizing their interfaces.

Between the design primitive classes, formal relations between the design representations are established: behavioral terms $H_1, ..., H_n$ *models* an aspect r, a behavioral term H_k may be derived from a set of behavioral terms $H_1, ..., H_n$ by applying a *transformation* rule, an aspect r_ν *refines* one or more behavioral terms H_ν, and an aspect r_k is hierarchical *decomposed* by a set of aspects $r_1, ..., r_m$.

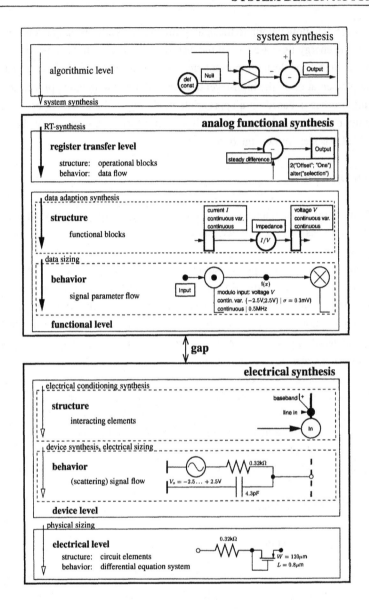

Fig. 1 Design flow

On this basis, the transition from a higher to a lower level of abstraction is described. The design knowledge is given by design primitives, and by design algorithms, specified by the formal relations between the design representations.

3.2 Information Domain Design Representation

From the structure point of view, the information domain is characterized by functional blocks representing principles of implementation. A principle of implementation is designated by specific characterizations of the data regarding the signal parameter that carry information, the value domain, and the characteristics of the events representing data validity in the time domain. Whereas this level of abstraction enables well known high level design methodologies, the principles of implementation consider the capabilities of analog systems.

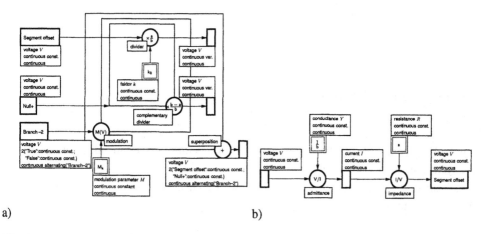

a)   b)

Fig. 2 a) Changeover switch decomposition, b) divider decomposition.

Fig. 2 a) shows the functional blocks of a changeover switch, derived by the analog functional synthesis of a modulo operator [21]. The divider is further decomposed into the fundamental principles of implementation admittance and impedance, as it is shown in Fig. 2 b). Quantitative, e. g. behavioral models on the functional level of abstraction ascertain the modulation of the signal parameters in the functional domain by means of a directed flow graph. This behavioral model is established from the functional block structure according to the *model* relation, and is called signal parameter flow, here.

In the signal parameter flow, parameter sources ($\otimes$) and drains ($\odot$) are assigned to the inputs and outputs of each functional block. These sources and drains quantitatively represent the properties of the corresponding signal parameters. This way, the links between the functional blocks are modeled as functions and are provided with a tolerance model, enabling freedom for the electrical domain synthesis steps.

As the result of the sizing process, detailed characterizations can be derived for the signal parameters. These characterizations may include the modulation ranges of the information carriers, their tolerance bounds by error limits Δ, or by variance σ, their bandwidth, and, if necessary, the sampling rate.

Besides the functional terms describing the principles of implementation, the signal parameter characterization contains tolerance terms regarding the linearity error, the offset value, and

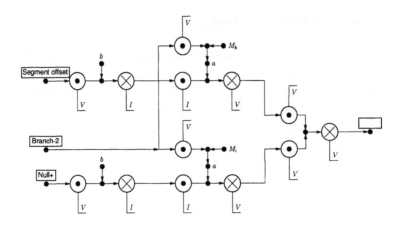

Fig. 3 Signal parameter flow of a changeover switch

noise. All tolerance terms refer to the signal parameters that carry information with respect to the system function, exclusively.

Fig. 3 shows the flattened signal parameter flow of a changeover switch as it is used for the modulo operator [22].

3.3 Electrical Domain Design Representation

In the electrical domain, the design is given by a structure of electronic devices, and, at a higher level of hierarchy, by blocks of devices.

The behavior of the design is derived from the device primitives behavioral models by means of the nodal analysis, for example. Beside application independent electrical models, a higher level of abstraction is enabled by using functional models for the device primitives. Due to the facts that these functional models are expressed by node voltages and branch currents, and they give a non-ambiguous description of the electric circuit from the electric power flow point of view, this level of abstraction belongs to the electrical domain, and is called device level, here.

Again, the design knowledge in electrical domain may be represented by formal relations. The behavioral descriptions are linked to the device primitives by the *model* relation. The nodal analysis gives *transformation* rules for building the behavioral description of the design from the behavioral descriptions of the device primitives.

The electrical domain behavioral descriptions may be represented by means of directed signal flow graphs, the so called MASON graphs [23][24]. The transfer function is given by the MASON formula, depending on the input and output specification. Functional models provide simplified graphs by focusing to special aspects. For example, the usage of ac models eliminates the operating point feeding. The two-way connected, directed graph expresses the strong

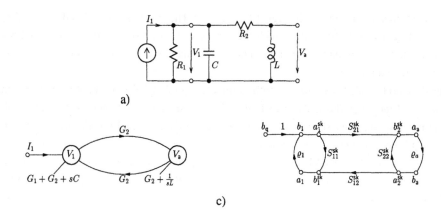

Fig. 4 Resonant circuit example: a) device structure, b) signal flow, c) scattering flow.

interaction between the devices of analog circuits. For example, Fig. 4 b) shows the signal flow of the resonant circuit in Fig. 4 a).

A scattering parameter based behavioral description gives the directed power flow in the electric circuit [25]. The wave parameters a and b represent the in- and outgoing voltage waves V^+ and V^- according to

$$a_n = \frac{V_n^+}{\sqrt{R_n}} \quad \text{and} \quad b_n = \frac{V_n^-}{\sqrt{R_n}}$$

with R_n denoting the characteristic impedance. Fig. 4 c) shows the scattering flow of the resonant circuit. The Reflection factors are are known to be

$$\varrho = \frac{R - R_n}{R + R_n}.$$

The wave parameters may be derived in a formal step from the node voltages and branch currents and vice versa by

$$I_n = \frac{a_n - b_n}{\sqrt{R_n}} \quad \text{and} \quad V_n = \sqrt{R_n}(a_n + b_n).$$

On this basis, the electrical signal may easily be decomposed in orthogonal components, represented by power channels. The scattering parameter based model may be represented by a MASON graph, too.

4 Information to Electrical Domain Mapping

4.1 Basic Approach

In the basic approach [18], the domain transition is done by establishing information domain models for the electrical domain blocks.

The modeling is done by a analysis approach. Starting with a signal flow graph representing the behavior of the electrical block, simplifications are done by symbolic methods, detecting the dominant interactions within the signal flow graph. The result is a simplified, directed graph representing a functional model. With the supposition to let only node voltages and branch currents be the signal parameters representing information, this graph is comparable with the signal parameter flow graph.

The structural synthesis step starts from a directed graph specifying the information domain behavior of the target design. For a successful synthesis, the specification is formal simplified by transformation rules, such as absorbing zero, neutral element (eliminate $\times$ 1), complementary sequence, complementary alternative, associative sequence, associative alternative, forward factorization, and backward factorization. The electrical domain structural blocks are identified on the basis of their functional domain models by a pattern recognition approach, for example. The next synthesis step may be done by hierarchical decomposition in the electrical domain.

4.2   Expanded Approach

Here, the information may be represented in the signal parameter flow by any signal parameter, including phase, frequency, level, pulse width, etc. Again, modeling the electrical domain design primitives in the information domain gives the key to the domain transition.

On the device level, all available device principles are modeled by scattering parameter based signal flow graphs. Every distinguishable, orthogonal signal component, relevant for the device principle model from the functional point of view, is assigned to a separate port, called *interaction channel*. In this way, physical signals are disassembled into signal components. The decomposition of electrical signals into orthogonal power flow channels gives freedom to the next synthesis phase, since the allocation of the power flow channels to electrical nodes is kept open.

In the next step, the signal flow graph is simplified with respect to idealized reflection coefficients. Only three cases have to be considered:

The error free propagation of a voltage signal from a source to a load port requires an open circuit impedance relation. This corresponds to a source reflection factor $\varrho_q = -1$ and a load reflection factor $\varrho_l = +1$. Assuming neglectable transmition line effects, the signal flow graph is simplified by loop elimination as it is shown in Fig. 5 b).

For the error free propagation of a branch current signal a short circuit impedance relation is needed. Here, the source reflection factor is $\varrho_q = +1$ and the load reflection factor is $\varrho_l = -1$. The simplified interconnection model is shown in Fig. 5 c).

In some cases, the propagation of a power signal is required. This is achieved by a matched impedance relation, giving both the source reflection factor ϱ_q and the load reflection factor $\varrho_l = 0$. Fig. 5 d) shows the simplified interconnection model.

From the functional point of view, an ohmic admittance may be used as a voltage-current converter with an open circuit impedance relation at its virtual input port, and a short circuit impedance relation at its output port. Both the input and the output ports belong to the same physical port. Fig. 6 shows the simplified, functional scattering signal flow graph. This interaction channel based electrical model may easily be related to the information flow as it is shown in Fig. 3 by formal transformation according to the relations between the signal parameters

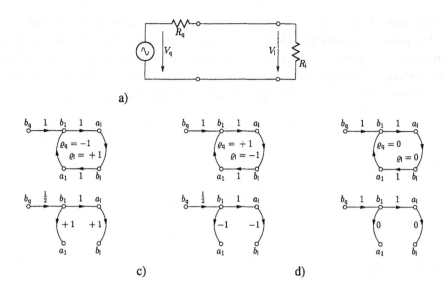

a)

b)   c)   d)

Fig. 5 Connection model simplification: a) circuit model, b) open circuit, c) short circuit, d) matched circuit.

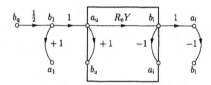

Fig. 6 Simplified scattering signal flow of a ohmic admittance

carrying information and the electrical signals.

On the basis of this simplified model, the connectability of several device principles depends on the interaction channel impedance characterization. Up to now, there is no quantitative parameter available, since they depend on the still unknown circuit structure. The impedance relations are characterized by open $(+)$, short $(-)$ or matched (0) relation.

In order to build up a circuit structure by connecting device principles, some more characterizations of the interaction channels are required.

The line mode characterization describes the physical nodes associated with the interaction channel to be driven in a balanced (double flag) or in a unbalanced mode (single flag).

The signal mode characterization describes the frequency range (operation point op, quasi-static qst, baseband bb, sideband sb, ...), or the time slot the signal belongs to.

On the basis of this characterizations, parasitic signal components are associate to the interaction channels. In this way, an interaction cannel consists of a primary signal component according to the device principles function, and several signal components with orthogonal characterizations specifying functional error. By connecting device principles with conformable characterizations of the primary signal components, a functional error may arise from parasitic signal components with equivalent characterizations. In the example of the ohmic admittance,

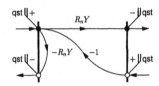

Fig. 7 Interaction channel model of a ohmic admittance

there are two parasitic signal components describing the error caused by the input current, and by the output voltage, respectively. In Fig. 7 the primary components are marked by solid dots in the interaction model. The parasitic interactions may be suppressed by either inserting an insulating device principle or by applying a transformation rule.

In the case of orthogonality with any other interaction channel, the operation point of a device principle is adjusted via separate interaction channels in the operation-point signal mode (op). In parallel to the information oriented signal flow, the bias feedings are structured in the same way.

On the basis of these fundamentals, the structural synthesis step is done by identifying the device principles from the signal parameter flow by a tolerant pattern recognition approach, and by applying transformation rules. The device level structure is built up by connecting the device principle primitives according to their interface characteristics.

In the first synthesis step, the dividers admittance and impedance principles of implementation within the changeover switch are assigned to the device principles *MOS-channel admittance* and *modulated MOS-channel impedance*. As it is shown in Fig. 8, this leads to a parasitic interaction loop. This parasitic interaction loop is eliminated by a device splitting transformation, resulting in a corrected admittance parameter

$$Y'_{\mathrm{DS}} = \frac{Y_{\mathrm{DS}}}{1 - R_{\mathrm{DS}}Y_{\mathrm{DS}}}.$$

The structure of electrical devices is derived from the interaction model by applying any of the following device synthesis transformations:

- parallel combination

- serial combination

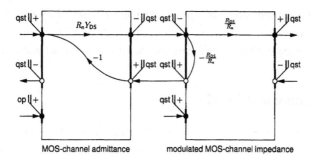

MOS-channel admittance   modulated MOS-channel impedance

Fig. 8 Parasitic interaction loop

- parallel connection

- serial connection

- channel combination

- serial splitting

- parallel splitting

- channel inversion

The changeover switch is built up by applying a series connection transformation, channel combination and channel inversion transformations, resulting in the circuit shown in Fig. 9.

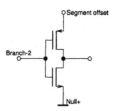

Fig. 9 Final device structure of the changeover switch

Compared with the basic approach, the electrical domain elements are of less complexity, since only functional aspects are to be considered in the electrical domain. Interface adaptions, signal conditionings, and operating point feedings are taken into account by separate device principles in the electrical domain.

The expanded approach enables formal descriptions of more abstract and elementary device principles.

5 Implementation and Results

The approach introduced here is implemented by Mathematica packages, providing an experimental system. Several parts, such as the design graph data base and pattern recognition engine, and the design graph visualization, are implemented by external C- and Java-routines, and are linked to Mathematica via the MathLink interface.

Within this system, the formal description of the synthesis methodology, and a general abstraction schema enables the usage of different approaches within the same design.

Up to now, the approach introduced here is demonstrated by two applications. The analog implementation of a modulo operator uses node voltages and branch currents to represent information. The quadrature demodulator sensor front end additionally requires phase and frequency parameters in the information domain.

The key problem of this design methodology is given by the design primitive modeling task.

Literature

[1] P. E. Allen and P. R. Barton. A silicon compiler for successive approximation A/D and D/A converters. In *Proc. IEEE Custom Integrated Circuits Conf.*, pages 552–555, 1986.

[2] M. G. R. Degrauwe et al. IDAC: An interactive design tool for analog cmos circuits. *IEEE J. Solid-State Circuits*, 22(6):1106–1115, Dec. 1987.

[3] J. G. Kenney and L. R. Carley. CLANS: A high-level synthesis tool for high resolution data converters. In *Proc. IEEE Int. Conf. Comput. Aided Design*, pages 496–499, 1988.

[4] E. Berkcan, M. d'Abreu, and W. Laughton. Analog compilation based on successive decompositions. In *Proc. ACM/IEEE-CS Design Automation Conference*, pages 369–375, 1988.

[5] R. Harjani, R. A. Rutenbar, and L. R. Carley. Oasys: A framework for analog circuit synthesis. *IEEE Trans. Computer Aided Design*, 8(12):1247–1265, Dec. 1989.

[6] El Turky Fatehy and E. E. Perry. BLADES: An artificial intelligence aproach to analog circuit design. *IEEE Transaction on Computer-Aided Design*, 8(6):680–691, June 1989.

[7] A. H. Fung, B. W. Lee, and B. J. Sheu. Self-reconstructing technique for expert system-based analog ic designs. *IEEE Trans. on Circuits and Systems*, 36(2):318–319, Feb. 1989.

[8] G. Jusuf, P. R. Gray, and A. L. Sangiovanni-Vincentelli. CADICS-Cyclic analog-to-digital converter synthesis. In *Proc. IEEE Int. Conf. Comput. Aided Design*, pages 286–289, 1990.

[9] S. G. Sabiro, P. Senn, and M. S. Tawfik. HiFADiCC: A prototype framework of a highly flexible analog to digital converter silicn compiler. In *Proc. IEEE Int. Symp. Circuits Syst.*, pages 1114–1117, 1990.

[10] G. F. M. Beenker, J. D. Conway, G. Schrooten, and A. G. J. Slenter. Analog CAD for consumer ICs. *Proc. Advances in Analog Circuit Design*, 1992.

[11] H. Chang and A. Sangiovanni-Vincentelli et al. A top-down, constraint-driven design methodology for analog integrated circuits. In *IEEE CICC*, May 1992.

[12] E. S. Ochotta, R. L. Carley, and R. A. Rutenbar. ASTRX/OBLX: Tools for Rapid Synthesis of High Performance analog Circuits. In *ACM/IEEE DAC*, pages 24–30, June 1994.

[13] W. Nye, D. C. Riley, A. Sangiovanni-Vincentelli, and A. L. Tits. DELIGHT.SPICE: An Optimization-Based System for the Design of Integrated Circuits. *IEEE Transactions on Computer-Aided Design*, 7(4):501–519, April 1988.

[14] F. Medeiro, F. V. Fernandez, R. Dominguez-Castro, and A. Rodriguez-Vazquez. A Statistical Optimization-Based Approach for Automated Sizing of Analog Cells. In *IEEE/ACM International Conference on CAD*, pages 594–597, 1994.

[15] H. Onodera, H. Kanbara, and K. Tamaru. Operational amplifier compilation with performance optimization. *IEEE J. Solid-State Circuits*, 25(2):446–473, Apr. 1990.

[16] G. G. E. Gielen, H. C. C. Walscharts, and W. M. C. Sansen. ISAAC: A symbolic simulator for analog integrated circuits. *IEEE J. Solid-State Circuits*, 24(6):1587–1596, Dec. 1989.

[17] J. P. Harvey, I. E. Mohamed, and B. Leung. STAIC: An Interactive framework for Synthesizing CMOS and BiCMOS Analog Circuits. *IEEE Transactions on Computer-Aided Design*, 11(11):1402–1417, November 1992.

[18] N. C. Horta and J. E. Franca. Automatic synthesis of data conversion systems using symbolic techniques. *Proc. IEEE Midwest Symp. Circuits Systems*, pages 877–880, 1995.

[19] C. Grimm, F. Heuschen, and K. Waldschmidt. Top-down design of mixed-signal systems with kandis. In *Workshop on System Design Automation SDA98*, pages 161–167, Dresden University of Technology, Dresden, March 1998.

[20] J. Kampe. A new methodology for the structural synthesis of analog blocks. In *IONS'98 Implementation Of spiking Neural Systems*, TU Ilmenau, Nov. 1998.

[21] J. Kampe. High level modeling of an analog modulo operator. In *IONS'98 Implementation Of spiking Neural Systems*, TU Ilmenau, Nov. 1998.

[22] J. Hichert and J. Kampe. A global optimization method and its application to the sizing of analog functional blocks. In *Proc. International Workshop on Global Optimization Go . 99*, Firenze (Italy), Sept. 1999.

[23] S. J. Mason. Feedback theory — some properties of signal flow graphs. *Proc. I. R. E.*, 41(9):1144–1156, Sept. 1953.

[24] S. J. Mason. Feedback theory — further properties of signal flow graphs. *Proc. I. R. E.*, 44(7):920–926, July 1956.

[25] K. C. Gupta, R. Garg, and R. Chadha. *Computer-Aided Design of Microwave Circuits*. Artech House Inc., Canton, 1981.

Statistical Analysis of Analogue Structures

A. Graupner, W. Schwarz, K. Lemke, S. Getzlaff, R. Schüffny,
Institut für Grundlagen der Elektrotechnik und Elektronik
Fakultät Elektrotechnik
Technische Universität Dresden

Abstract

This paper treats the tolerance analysis of analogue systems with the objective of statistical modelling of complex analogue systems. A methodology using variance calculation is proposed and compared with Monte-Carlo simulation in terms of effort and accuracy. It is shown that for variance calculation a linear approxiation is sufficient for typical analog circuits and that it requires less effort than Monte-Carlo analysis while providing comparable or even better accuracy.

1   Introduction

The fabrication of heterogeneous systems in general involves various manufacturing steps. Due to fabrication principles the behaviour of the elements the system comprises of can randomly deviate from their nominal. For example, the production of integrated circuits incorporates stochastical processes like implantation and therefore, the parameters of the produced elements are subject to random variations. Such effects in general limit the performance of the whole system and especially of the continuous-valued or analogue blocks. Thus, for robust system and circuit design it is crucial to investigate the influence of parameter fluctuations.

Once having statistical element models it is possible to estimate the accuracy of building blocks or the manufacturing yield. A frequently used methodology is Monte-Carlo analysis [1]. Accomplishing a Monte-Carlo analysis is very computing power and time consuming as element-level simulations have to be performed many times. In order to simplify tolerance analysis a hierarchical sensitivity analysis and variance calculation is proposed to evaluate the influence of elemental parameter variation [2].

This chapter is intended to enlight these problems from the analogue circuit designers viewpoint. After the problem statement we explain the methodology of variance calculation for the analysis of the influence of device mismatch on circuit performance as an alternative to Monte-Carlo analysis. It is shown that for the general cases linearized variance analysis is sufficient in terms of accuracy. After reviewing a simple statistical model of a MOS-transistor our methodology is illustrated at the example of a simple transconductance amplifier circuit and compared with the results of a Monte-Carlo analysis.

2   Problem statement

The precision of analogue circuits in general depends on the accurate matching of devices, e. g. that two or more devices behave indentically. But mismatch between them can exist due to random device parameter variation. As a rule random parameters are assumed to be normally distributed. This also holds true for the output signal of a system if it is sufficiently linear over the random parameter variation interval and if it comprises of a large number of random parameters. Thus, the deviation of the output signal from the nominal in terms of a standard deviation σ can be used as a measure. It can be determined by means of Monte-Carlo simulation or variance calculation as explained in the following sections. Our objective is to derive statistical models of the analogue circuits for system simulations (figure 1) and basing

R. Merker and W. Schwarz (eds.), System Design Automation, 164-175.

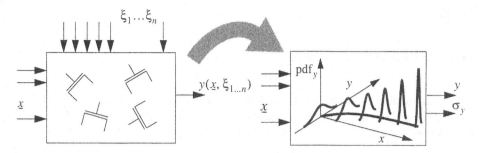

Fig. 1 (left) A transistor level model has a large number of input whereas (right) the behavioural model is much more simple.

thereupon a statistcal model for the whole system.

3   Tolerance analysis

The analysis of the impact of device mismatch in analog circuits in general is rather complicated. Often the circuits comprise of a large number of devices and the mismatch-contribution of any of them has to be taken into account. Furthermore, as a rule the influence of the device mismatch on the circuit's transfer function is of a non-linear nature. Hence, a widespread used methodology for analysing device mismatch in analog circuits is Monte-Carlo analysis [1, 5-9].

In a Monte-Carlo analysis a simulation is repeated several times (100-1000) with pseudo-random model parameter sets. From the results of these simulations mean value and variance are extracted. Thus, the influence of device mismatch can be estimated and eventually, a statistical model of the analysed circuit can be derived. The accuracy of this method depends on the number of samples, i. e. the number of performed simulations. As a large number of simulations have to be carried out a Monte-Carlo analysis is very time consuming. Alternatively tolerance analysis can be accomplished analytically.

3.1   Variance calculation

Figure 2 shows a parallel system comprising of a number of functional blocks. The outputs η_v of all function blocks g_v is combined to a single output value η by the function G.

A single function block of the parallel structure is depicted in figure 3. Its output is a random

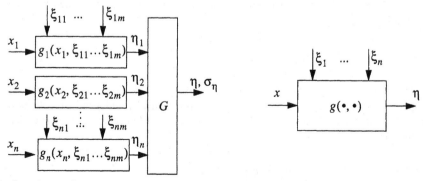

Fig. 2   Array of processing elements g_v **Fig. 3**   Function block

variable

$$\eta = g(x, \underline{\xi}) \tag{1}$$

where x is the deterministic input and

$$\underline{\xi} = (\xi_1, \xi_2, ..., \xi_n) \tag{2}$$

is the vector of random parameters. The function g can be chosen such that the ξ_i $(i = 1, ..., n)$ are zero-mean. The problem of statistical analysis consists in determining the statistical properties of η, especially those of the random deviation from the nominal value $g(x, \underline{0})$.

$$\delta = \eta - y = g(x, \underline{\xi}) - g(x, \underline{0}) \tag{3}$$

The corresponding general error model is shown in figure 4.

The calculation of the statistical characteristics of δ is complicated in general. However, in most practical cases the function g is only weakly nonlinear within the variation intervals of the ξ_i. Then the linear approximation

$$g(x, \underline{\xi}) = g(x, \underline{0}) + \sum_{i=1}^{n} a_i \xi_i = g(x, \underline{0}) + \delta, \tag{4}$$

with sensitivity coefficients a_i [3] given by

$$a_i = a_i(x) = \left. \frac{\partial}{\partial \xi_i} g(x, \underline{\xi}) \right|_{\underline{\xi} = \underline{0}} \tag{5}$$

suffices. This simplifies the statistical analysis substantially. Eq. (4) corresponds to the linear error model depicted in figure 5.

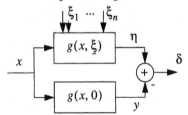

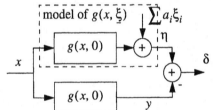

Fig. 4   Error model of the function block

Fig. 5   Linear error model of the function block

Here due to (4) the stochastic influences of all random parameters are modelled by a single random variable

$$\delta = \sum_{i=1}^{n} a_i \xi_i \tag{6}$$

describing the random deviation from the desired output value $g(x, 0)$ at a given input value x. If as generally assumed the ξ_i $(i = 1, ..., n)$ are stochastically independent with normal distribution $N(0, \sigma_i^2)$ then δ as well as η have a normal distribution too and their variances calculate from (6) as

$$\sigma_\delta^2 = \sigma_\eta^2 = \sum_{i=1}^{n} a_i^2 \sigma_i^2 \tag{7}$$

Thus in order to determine the variance of the output η in (3) only the coefficients a_i have to be provided. They can be calculated analytically using (5), derived by means of small signal or sensitivity analysis, taken from the Jacobian matrix at the bias point [10] or computed by numerical differentiation with

$$a_i \approx \frac{g(x, 0, ..., \Delta, ..., 0) - g(x, 0, ..., 0, ..., 0)}{\Delta} \qquad (8)$$

using an appropriate step value Δ for each variable. In the latter case for the calculation of σ_δ^2 $n+1$ transfer functions have to be determined. This method also can be utilized in discrete time systems where other approaches may fail. Note that, according to (5) the a_i depend on the input x.

3.2 Accuracy considerations

In order to demonstrate how the nonlinearity of g effects the result (7) we regard the simple system

$$\eta = g(\xi) = a(x)\xi + b(x)\xi^2 = a\xi + b\xi^2, \qquad (9)$$

where ξ is a $N(0, \sigma_\xi)$ distributed random variable. Here the variance of η calculates as

$$\sigma_\eta^2 = a^2\sigma_\xi^2 + 2b^2\sigma_\xi^4 \qquad (10)$$

and the relative error in comparsion to (7) will be

$$\frac{\Delta\sigma_\eta^2}{\sigma_\eta^2} = 2z^2 \qquad \text{with} \qquad z = \frac{b\sigma_\xi}{a} \approx b\sigma_\eta. \qquad (11)$$

From this with given coefficients a and b it can be estimated, to what extend ξ resp. η may vary (σ_ξ resp. σ_η) so that the error of (7) does not exceed a certain limit.
The deviation of the probability density function $f_\eta(y)$ of η from a normal distribution can be estimated by using the coefficients of

skewness and excess

$$\gamma_{\eta 3} = \frac{\mu_{\eta 3}}{\mu_{\eta 2}^{3/2}} \qquad \text{and} \qquad \gamma_{\eta 4} = \frac{\mu_{\eta 4}}{\mu_{\eta 2}^2} - 3 \qquad (12)$$

where $\mu_{\eta n}$ are the n-th order central moments of the distribution $f_\eta(y)$. Note that third and higher order coefficients are zero for a normal distribution. A calculation of the moments of (9) leads to

$$\gamma_{\eta 3} = 2z\frac{3 + 2z^2}{(1 + z)^{3/2}} \qquad \text{and} \qquad \gamma_{\eta 4} = 48z^2\frac{1 + z^2}{(1 + 2z^2)^2} \qquad (13)$$

with z from (11). Figure 6 shows the dependence of these characterisics from z. As a rule values of $|\gamma_{\eta 3}| < 0.1$ and $|\gamma_{\eta 4}| < 0.1$ are sufficient for $f_\eta(y)$ to be close to normal. This leads to $|z| < 0.05$ and consequently $\sigma_\xi < 0.05|a/b|$ resp. $\sigma_\eta < 0.05/|b|$ as reasonable estimates for the tolerable standard deviations of ξ resp. η. As a demonstrative example figure 7 shows $f_\eta(y)$ (bold line) for $z = 0.05$ in comparsion with a normal distribution (thin line).

4 Hierarchical tolerance analysis

Once having a statistical description of the function blocks, the above method can be used hierarchically until the whole system is characterized:
The transfer functions of the function blocks are extracted from transistor level circuits and the

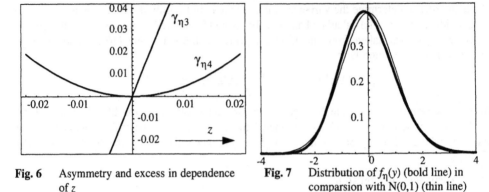

Fig. 6 Asymmetry and excess in dependence of z

Fig. 7 Distribution of $f_\eta(y)$ (bold line) in comparsion with N(0,1) (thin line)

standard deviation of the output signal is derived as a function of the input signals. Thus, a behavioural model consists of two parts: the mean value of the transfer function and the signal dependend standard deviation (figure 1). These behavioural models are used for simulation in the next hierarchy level (figure 8). The problem to solve is the same as above: a block with several deterministic and random inputs has to be characterized. Thus, for statistical analysis the linear variance calculation can be employed here as well with the same arguments given in the previous chapter. Even more, if the number of function blocks is large the output signal will become normally distributed with very good approximation. Consequently, for obtaining the standard deviation σ of the output signal y for a given input value x as much derivatives (a_i) have to be determined as the number of function blocks the hierarchy level comprises of (figure 8).

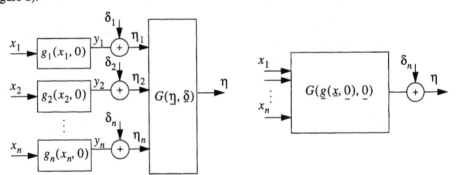

Fig. 8 Hierarchical statistical characterisation of the system. All random variables are combined to a single value.

5   Comparison with the Monte-Carlo analysis

In Monte-Carlo analysis an ensemble of transfer curves is calculated from which the statistical characteristics are estimated. In order to compare the effort of variance calculation with Monte-Carlo analysis we determine the sample size necessary for a given accuracy, i. e. in statistical terms confidence level and confidence interval. From estimation theory it is known, that for a sufficently large sample size n the estimate

$$\hat{m} = \frac{1}{n} \cdot \sum_{i=1}^{n} \eta_i \qquad (14)$$

of the mean $m = E(\eta)$ has a $N(m, \sigma^2/n)$ distribution, where σ^2 is the variance of η. Thus, with confidence level γ it lies within the interval

$$m - c\frac{\sigma}{\sqrt{n}} \leq \hat{m} \leq m + c\frac{\sigma}{\sqrt{n}}, \tag{15}$$

where c is given by the interval probability

$$P(-c \leq \zeta \leq c) = 2 \cdot \text{erfc}(c) - 1 = \gamma \tag{16}$$

of a $N(0,1)$ distributed random variable ζ. For the typical values $\gamma = 0.95$ and $\gamma = 0.99$ we have $c = 1.96$ and $c = 2.58$ respectively. From this with given interval width

$$\Delta\hat{m} = 2c\frac{\sigma}{\sqrt{n}} \tag{17}$$

we obtain the necessary sample size n as

$$n = 4 \cdot \left(c\frac{\sigma}{\Delta\hat{m}}\right)^2. \tag{18}$$

If σ is not known, for sufficiently large sample sizes the estimate $\hat{\sigma}$ from below can be used. Similar to that, the estimate

$$\hat{\sigma}^2 = \frac{1}{n-1}\sum_{i=1}^{n}(\eta_i - m)^2 \tag{19}$$

of the variance has a $N(\sigma^2, 2\sigma^4/n)$ distribution. We want the estimate (19) with probability γ to fall into the interval

$$\sigma^2 - c\sqrt{\frac{2}{n}}\sigma^2 \leq \hat{\sigma}^2 \leq \sigma^2 + c\sqrt{\frac{2}{n}}\sigma^2. \tag{20}$$

The width

$$\Delta\hat{\sigma}^2 = 2c\sqrt{\frac{2}{n}}\sigma^2 \tag{21}$$

of the variance interval corresponds to the interval width

$$\Delta\hat{\sigma} = \Delta\hat{\sigma}^2/2\sigma \tag{22}$$

of the standard deviation. From that the required sample size calculates to

$$n = 2 \cdot \left(c\frac{\sigma}{\Delta\hat{\sigma}}\right)^2. \tag{23}$$

Some numerical examples may be found in table 1.

We note that in a Monte-Carlo analysis the accuracy of the estimation depends on the number of samples only and thus, is independent of the number of parameters subject to random variation [1]. Especially in very large systems with a flat hierarchy a Monte Carlo analysis would require less simulations for achieving a comparable accuracy than variance calculation. A simple example of a typical function block is analysed in the next section.

Table 1: Reqiured number of samples for estimation of mean and standard deviation with a confidence level of 0.99 ($c \approx 2.5$) and given accuracy

accuracy of the estimate $\dfrac{\Delta \hat{m}}{\sigma}$ resp. $\dfrac{\Delta \hat{\sigma}}{\sigma}$	Required number of samples for estimation of	
	mean	standard deviation
25%	400	200
10%	2,500	1,250
5%	10,000	5,000
1%	250,000	125,000

6 Analysis example

6.1 Transistor Mismatch Model for Weak Inversion

The drain current of a saturated transistor biased in subthreshold with bulk shorted to source is

$$I_D = I_{D0} \cdot \exp\left(\frac{V_{GS} - V_t}{nU_T}\right) \tag{24}$$

where I_{D0} is a current constant, V_t is the threshold voltage and nU_T is the subthreshold slope [20]. The spatial fluctuations of the physical device parameters are the most important source of mismatch. Factors contributing to these fluctuations among others are variations of the effective mobility, variations in the oxide thickness and variations of the dopant-distribution in the subtrate. These variations can be modelled as a certain level of random mismatch in the device model parameters: as deviation of the current constant I_{D0} and the threshold voltage V_t from their nominals [11-13]. As reported in [14, 15] the variation of the subthreshold slope is negligible small. The random deviations of the threshold voltage ΔV_t and the current constant ΔI_{D0} are assumed to be normal distributed with zero mean and given standard deviations. The latter depends on the process and the device geometry, for non-minimum sized devices holds [11-13]

$$\sigma^2_{(\Delta Vt)} = \frac{A^2_{Vt}}{2WL} \qquad \sigma^2_{(\Delta ID0/ID0)} = \frac{A^2_\beta}{2WL} \tag{25}$$

where A_{Vt} and A_β are process constants and W and L the transistor's dimensions. These dependencies are investigated more accurately in [16-18].

Thus, considering mismatch for the drain current holds

$$I_D = (\overline{I_{D0}} + \Delta I_{D0}) \cdot \exp\left(\frac{V_G - (\overline{V}_t + \Delta V_t)}{nU_T}\right)$$

$$= I_{D0} \cdot \exp\left(\frac{V_G - \overline{V}_t - \Delta V_t + nU_T \ln(1 + (\Delta I_{D0}/\overline{I_{D0}}))}{nU_T}\right) \tag{26}$$

With

$$v_t = -\Delta V_t + nU_T \ln(1 + (\Delta I_{D0}/\overline{I_{D0}})) \tag{27}$$

eq. (26) can be written as

$$I_D = I_{D0} \cdot \exp\left(\frac{V_G - V_t - v_t}{nU_T}\right). \tag{28}$$

The v_t is zero mean and can be assumed to be normal distributed as $\Delta I_{D0}/\overline{I_{D0}}$ is small in general. The standard deviation can be computed from the process constants and device geometries (25):

$$\sigma_{vt}^2 = \sigma_{(\Delta Vt)}^2 + (nU_T \ln(1 + \sigma_{(\Delta ID0/ID0)}^2))^2. \tag{29}$$

So, the transistor parameter variations can simply be modelled with a voltage source with random value v_t which is connected in series with the gate.

6.2 Circuit Analysis

As an example a simple transconductance amplifier circuit will be analysed. The circuit diagramm is shown in figure 9. Using the above transistor mismatch model yields the circuit diagram depicted in figure 10.

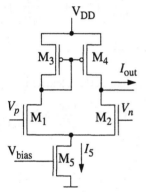

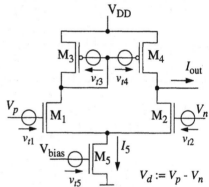

Fig. 9 Example circuit: a transconductance amplifier

Fig. 10 Transconductance amplifier: the parameter mismatch is modelled with voltage sources $v_{t1..5}$

The sensitivity coefficients a_i of eq. (5) are most effectively calculated by a small signal analysis. The corresponding small-signal equivalent circuit is shown in figure 11. An analysis of this circuit results in

$$i_{out} = \frac{2g_{m1}g_{m2}}{g_{m1} + g_{m2}}(v_{t2} - v_{t1}) + g_{m4}(v_{t4} - v_{t3}) + \frac{g_{m2} - g_{m1}}{g_{m1} + g_{m2}}g_{m5}v_{t5}. \tag{30}$$

From that the coefficients a_i immediately can be obtained as

$$-a_1 = a_2 = \frac{2g_{m1}g_{m2}}{g_{m1} + g_{m2}} \qquad -a_3 = a_4 = g_{m4} \qquad a_5 = \frac{g_{m2} - g_{m1}}{g_{m1} + g_{m2}}g_{m5}. \tag{31}$$

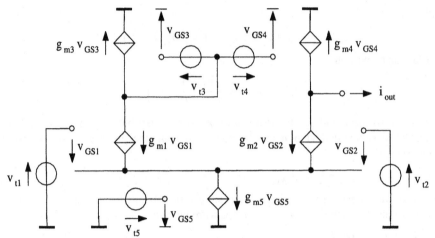

Fig. 11 Small signal equivalent circuit of the OTA (figure 10)

With the transconductances

$$g_{m1} = \frac{I_{D1}}{n_n U_T} = \frac{I_{50}}{2n_n U_T}\left(1 - \tanh\frac{V_d}{2n_n U_T}\right); \quad g_{m3} = \frac{I_{D3}}{n_p U_T} = \frac{I_{50}}{2n_p U_T}\left(1 - \tanh\frac{V_d}{2n_n U_T}\right) = g_{m4}$$

$$g_{m2} = \frac{I_{D2}}{n_n U_T} = \frac{I_{50}}{2n_n U_T}\left(1 + \tanh\frac{V_d}{2n_n U_T}\right); \quad g_{m5} = \frac{I_{D5}}{n_n U_T} = \frac{I_{50}}{n_n U_T}$$

$$(32)$$

the coefficients a_i in dependence from the difference input voltage V_d calculate as:

$$a_1(V_d) = \frac{I_{50}}{-2n_n U_T \cdot \cosh^2\left(\frac{V_d}{2n_n U_T}\right)} = -a_2(V_d) \qquad a_5(V_d) = -\frac{I_{50}}{n_n U_T}\tanh\frac{V_d}{2n_n U_T}$$

$$(33)$$

$$a_3(V_d) = \frac{I_{50}}{2n_p U_T}\left(\tanh\left(\frac{V_d}{2n_n U_T}\right) + 1\right) = -a_4(V_d)$$

For the numerical calculation the following parameter values have been used: $\sigma_{vt1} = \sigma_{vt2} = \sigma_{vt5} = 1.09$ mV, $\sigma_{vt3} = \sigma_{vt4} = 1.34$ mV, $n_p U_T = 31.3$ mV, $n_n U_T = 34.5$ mV and $I_{50} = 50$ pA. In figure 12 the coefficients a_i are depicted in dependence from the difference input voltage V_d.

As it is easily seen, in the circuit in figure 10 the influences of threshold-voltage deviations in the current mirror (M3, M4) and in the current transistor M5 are stronger than those in the difference pair (M1, M2).

According to (7) the standard deviation of the output current calculates as

$$\sigma_{Iout}(V_d) = \sqrt{\sum_{i=1}^{5}(a_i(V_d)\sigma_{vti})^2}$$

$$(34)$$

It is depicted in figure 13 (solid line).

For comparsion a Monte-Carlo simulation has been carried out. As a direct result, the sample of transfer functions is depicted in figure 14. From this the average transfer function (figure 15), the random deviations from the average (figure 16) and the standard deviation of the output

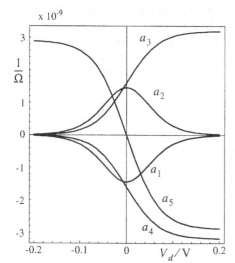

Fig. 12 Value of the coefficients a_i in dependence from V_d for the circuit of fig. 10

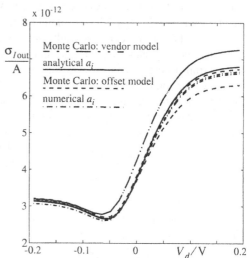

Fig. 13 Standard deviation of the transfer function obtained with various methods

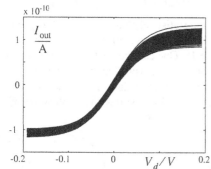

Fig. 14 Sample of transfer functions as direct result of a Monte-Carlo simulation

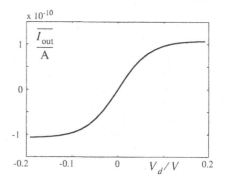

Fig. 15 Average transfer function from figure 14

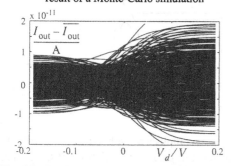

Fig. 16 Random deviation of the transfer function from its average

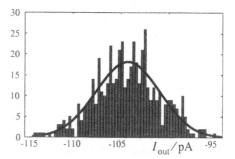

Fig. 17 Histogramm of the output current at V_d=0.14 V

current (figure 13) have been calculated. Furthermore a histogramm of the output current at V_d = -0.14 V is shown in figure 17 which provides an impression about how the random component of the output current is distributed. All data are based on a sample of 500 curves, which took approximately 2 hours simulation time to be obtained.

The curves of figure 13 represent results from Monte-Carlo anaylsis with the vendor's transistor mismatch model (vendor model) [19] and the mismatch model of section 6.1 (offset model). Results from a variance analysis with the coefficients a_i calculated analytically and numerically are shown for comparison.

The analytical results coincide completely with those obtained from circuit simulation. The curve obtained with the „real" vendor's transistor model differs slightly because the mismatch mechanism is modelled more accurately than in the offset model. The difference between the two curves obtained by the Monte-Carlo method with offset model result from the fact, that they represent estimates and thus realisations of random values. Note that the 0.99 confidence interval for a sample size of 500 is about $\Delta\sigma = 0.8 \cdot 10^{-12}$ A.

6.3 Accuracy estimation

For the circuit in figure 10 an explicit expression of the output current I_{out} in dependence from the input difference voltage $V_d = V_p - V_n$ and the mismatch parameters v_{ti} can be found:

$$I_{out}(V_d, v_{t1}, v_{t2}, v_{t3}, v_{t4}, v_{t5}) =$$

$$= \frac{I_{50}}{2}\exp\left(\frac{-v_{t5}}{n_n U_T}\right)\left(1 + \exp\left(\frac{v_{t3}-v_{t4}}{n_p U_T}\right)\right) \cdot \left[\tanh\left(\frac{v_{t3}-v_{t4}}{2n_p U_T}\right) + \tanh\left(\frac{V_d+v_{t2}-v_{t1}}{2n_n U_T}\right)\right] \quad (35)$$

So using the normal distribution of v_{ti} $(i = 1...5)$ the statistical characteristics of I_{out} can be calculated exactly. Following the treatment in sec. 3.2 the relative error of σ_{Iout} from (34) in comparsion with the exact result is depicted in figure 18. One can see, that this error is less than 0.35%. In figure 19 skewness and excess of the probability density function (pdf) of I_{out} are depicted over the difference input voltage. Comparing with the values in 3.2 one can see that the output pdf is very close to the normal distribution. This justifies the application of the linearized model to the variance analysis of the circuit example.

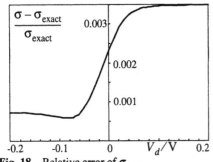

Fig. 18   Relative error of σ_{Iout}

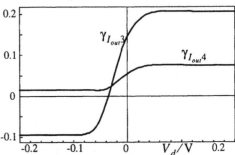

Fig. 19   Asymmetry and excess

7   Conclusion

Tolerance analysis of analogue systems can be accomplished with either Monte-Carlo simulation or variance calculation. The latter method can greatly be simplified if the analog elements are weakly nonlinear over the random parameter variation range because a small signal analysis can be applied. This is applicable for typical analogue circuits.

In order to determine the influence of element scattering on the system behaviour by Monte-Carlo analysis M simulations have to be carried out, where M is specified by the necessary accuracy of the estimation. In contrast for variance calculation N simulations have to be performed where N is the number of random parameters, e.g. elements.

Thus, in many cases tolerance analysis by means of variance calculation requires far less effort than a Monte-Carlo simulation with the same prescribed accuracy. Additionaly, with sensitivity analysis the designer gets a deeper insight how the parameter scattering influences the systems performance.

8   Acknowledgement

This work was partly supported by the Sonderforschungsbereich 358, Teilprojekte A7 and E1 of the Deutsche Forschungsgemeinschaft.

9   Literature

[1] R. Spence and R. S. Soin: Tolerance Design of Electronic Circuits, Imperial College Press, 1997.

[2] A. Graupner, W. Schwarz, K. Lemke, S. Getzlaff and R. Schüffny: Statistical Analysis of Parallel Analog Structures, System Design Automation SDA'2000, Rathen, Germany, March 2000.

[3] K. Reinschke and P. Schwarz: Empfindlichkeit dynamischer Systeme, in E. Phillipow: Taschenbuch Elektrotechnik, Band 2, pages 872-916. Verlag Technik Berlin, 3. Auflage, 1987.

[4] G. Debyser, F. Leyn G. Gielen, W. Sansen, and M. Styblinski: Efficient Statistical Analog IC Design Using Symbolic Methods. In Proc. ISCAS'1998.

[5] T. B. Tarim and M. Ismail. Enhanced Analog Yields Cost-Effective Systems-on-Chip. IEEE Circuits and Devices Magacine, 15(2):12-21, March 1999.

[6] T. B. Tarim, H. H. Kuntman, and M. Ismail. Robust Design of Basic Low Voltage CMOS Transconductors. Analog Integrated Circuits and Signal Processing, 22:175-191, 1999.

[7] Hing yan To, Hua Su, and Mohammed Ismail. A Novel Sensitivity Analysis Technique for VLSI Statistical Design. In Proc. ISCAS'96.

[8] K. Antreich, J. Eckmueller, H. Graeb, M. Pronath, F. Schenkel, R. Schwencker, and S. Zizala. WiKkeD: Analog Circuit Synthesis Incorporating Mismatch. In IEEE Custom Integrated Circuits Conference CICC'2000, pages 511-514, Orlando, Florida, May 2000.

[9] K. J. Antreich, H. E. Graeb, and R. K. Koblitz. Advanced Yield Optimization Techniques. In S. W. Director and W. Maly, Editors, Statistical Approach to VLSI, pp. 163-195. Elsevier Science, 1994.

[10] R. Lopez-Ahumadu and R. Rodriguez-Macias: Fastest: A Tool for a complete and efficient statistical Evalusation of analog Circuits, Proc. ISCAS'2000, Geneva, Switzerland, 2000.

[11] M. Pelgrom, A. Duinmaijer and A. Welbers: Matching Properties of MOS Transistors, IEEE Journal of Solid-State Circuits, Vol. 24, No. 5, 1989

[12] K. R. Lakshmikumar, R. A. Hadaway, and M. A. Copeland. Characterization and Modeling of Mismatch in MOS Transistors for Precision Analog Design. IEEE JSSC, 21(12):1057-1066, Dec. 1986

[13] C. Michael and M. Ismail. Statistical Modeling of Device Mismatch for Analog MOS Integrated Circuits. IEEE Journal of Solid-State Ciruits, 27(2):152-165, February 1992

[14] A. Pavasovic, A. Andreou and C. Westgate: Characterization of Subthreshold MOS Mismatch in Transistors for VLSI Systems, Analog Integrated Circuits and Signal Processing, Vol. 6, 1994.

[15] F. Forti and M. E. Wright. Measurement of MOS Current Mismatch in the Weak Inversion Region. IEEE Journal of Solid-State Circuits, 29(2):138-142, February 1994.

[16] S. J. Lovett, M. Wilten, A. Methewson, and B. Mason. Optimizing MOS Transistor Mismatch. IEEE Journal of Solid-State Ciruits, 33(1):147-150, January 1998.

[17] T. Serrano-Gotarredona and B. Linares-Barranco. Systematic Width and Length Dependent CMOS Transistor Mismatch Characterisation and Simulation. Analog Integrated Circuits and Signal Processing, 21:271-296, 1999.

[18] J. Oehm, U. Grünebaum, and K. Schumacher. A Physical Approach to Mismatch Modelling and Parameter Correlations. In Proc. ISCAS'2000 and ESSCIRC'1999.

[19] AMS: 0.6 μm CMOS CUP Process Parameters, 1998.

[20] C. Enz, F. Krummenacher and E. Vittoz: An Analytical MOS Transistor Model Valid in All Regions of Operation and Dedicated to Low-Voltage and Low-Current Applications, Analog Integrated Circuits and Signal Processing, 8(1):69-81, July 1995.

A New Hierarchical Simulator for Highly Parallel Analog Processor Arrays

Stephen Henker · Stefan Getzlaff · Achim Graupner
Jörg Schreiter · Mirko Puegner · René Schüffny

Stiftungsprofessur für Hochparallele VLSI-Systeme und Neuromikroelektronik
Institut für Grundlagen der Elektrotechnik und Elektronik
Fakultät Elektrotechnik
Technische Universität Dresden

Abstract

System simulation, evaluation and analysis of highly parallel analog systems with conventional simulation tools is quite difficult due to their high complexity. Therefore we developed a methodology for the analysis of such systems by means of an applied simulation system. This approach bases on a relaxation method. The application of this technique accelerates the simulation of highly parallel analog systems by several orders of magnitude compared with standard circuit simulators. As additional feature, it can be traded-off between simulation expenditure and accuracy. We present the simulation method and its advantages in addition to a practical example.

1 Introduction

High-performance data processing circuits with low power consumption are a technological key factor. Since conventional serial data processing designs reach their limits highly parallel systems are a promising alternative. Such systems consist of a large number of similarly structured elements. These elements are interconnected to a network that process data in parallel. The increasing parallelism relaxes the requirements of calculation speed of the single elements while providing competitive performance in comparison to sequential data processing.

In particular, parallel data processing is favorable if data is already present in parallel or can be parallelized easily. Image processing is one example of an area of application suited very well. An interesting field of these application are for instance CMOS-sensors with embedded data processing, like neuromorphic, vision and video-chips.

In order to implement parallel systems consisting of some 10.000 processing elements, a single element needs to be small in area besides consuming very little power. A promising realization can be found in analog domain. In analog circuits the properties of the semiconductor devices can be fully exploited and thus, very small functional elements can be developed.

Due to the large size of highly parallel analog systems simulation and verification is quite difficult. A single analog processor element can be simulated easily by means of standard circuit simulators. In contrast, simulation of the entire parallel processor array is very computing power and time consuming. Often, the use of standard circuit simulators is not practical due to their low simulation performance. It is caused by the employed simulation method: the solution of differential equations systems. For example, the most common used implicit intervening solution technique bases on the solution of a system of nonlinear differential equations. This is ac-

176

R. Merker and W. Schwarz (eds.), System Design Automation, 176-182.
© 2001 *Kluwer Academic Publishers.*

complished by solving several linear differential equations systems of same size until a solution with sufficient small error is found. Note, that the complexity of the equation system grows with the square of the number of components. Since parallel systems contain a large number of processor elements, the size and complexity of the equations system can easily exceed the abilities of solving methods implemented in standard circuit simulators. Thus, the practical feasibility of standard tools of the simulation of large systems strongly depends on their size.

Various approaches to simulate large analog systems exist, like down scaling or abstract mathematical approximation. Highly parallel systems are reduced to systems of lower complexity containing fewer elements that can be simulated with conventional tools. However this simplification usually assumes idealized and identical behavior of the processing elements.

But especially in analog systems the analysis of the non-ideal behavior is crucial. One critical point in the realization of analog systems is element mismatch. Mismatch describes the effect that identically designed devices on integrated circuits behave different due to the stochastic nature of physical processes in fabrication. Robust design methodologies have been developed to suppress the influence of device mismatch, i.e. switched current circuits as well as robust algorithms and system architectures specifically adapted to implementation. Anyway, these approaches need an appropriate method for verification too.

In order to accelerate the analysis of parallel analog systems a new methodology has been developed. We propose hierarchical scalable modeling in connection with a dedicated tool capable to simulate highly parallel analog systems fast and efficiently.

2 Simulator Functionality

The limited performance of standard circuit simulators when simulating large analog systems is mainly due to the implemented solving technique. As a rule the calculation expenditure is of the order $O[n^2]$ where n is the number of equations. Almost all commercial simulation tools use the same or a slightly modified method of solving nonlinear equations systems. Thus, their behavior is similar with respect to highly parall el analogsystems.

A new simulation system has been conceived which applies a method that is less computing intensive in order to accelerate the simulation of highly complex analog networks. For this aspecial relaxation method was chosen. It is used for modeling and simulation of networks on an abstract level.

The relaxation method itself is well known. So the usage of this method bases on profound results of scientific research [1]. Apart from well known properties the relaxation method requires with $O[n]$ substantially less calculation effort than methods used in standard circuit simulator. This is especially true for systems containing a large number of elements.

The underlying principle of this method is that the represented system can be described as a network consisting of interconnected data processing nodes. The node connections are unidirectional and describe the flow of information from a source to a sink, as illustrated in figure1.

A single network node represents a elementary processing unit of this simulation system. It processes input signals and feeds the according result to its output(s). The output signal can be used as input signal for further nodes. Information is processed on its way from node to node and a simulation can be done easily by calculating the state of each node at every time step.

Since relaxation method is iterative, there is the question if it converges to the correct solution and, if so, under what condition. These conditions are stated in theorems that can be found in references [2] and [3]. The current implementation uses a modified Gauss-Jakobi-Method for this iterative solution. It is computing efficient and is based on a localized view of the simulated system.

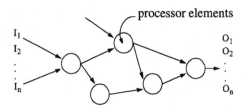

Fig. 1   Data Processing Network

Some disadvantages are inherent to relaxation methods. Due to the approximation of dynamic system behavior the relaxation method can not reach the accuracy of a waveform solution calculated from a differential equation system. But it is possible to minimize the systematic error by parameters of relaxation method. However, the main disadvantage of the relaxation method is that it is only fully suited for straightforward networks. Recursions within the information flow may cause the simulation method itself to behave instable. This is due to the delay caused by the step-wise node calculation. This delay has influences on the frequency response of the simulated system. Particularly systems with feedback structures become modified, poles of the frequency response function become changed or additional poles become created respectively. This can result in instabilities caused by the simulation method. There are different approaches to stabilize the simulation method in such cases, e.g. controlled signal attenuation.

However, in our special case the simulation environment is to be used for the simulation of straightforward networks where the advantages of relaxation method can be fully exploit.

3   Element modeling

In order to use relaxation method a specific representation of the system is needed. Each element of the system has to be represented by a node. These nodes are to be connected accordingly, in the same way as their counterparts in the original system. An important part of this representation is the transfer function of nodes. These functions model the behavior of the processing elements like adders, multipliers, etc. They have to be derived for each basic processing element individually. The modeling of the element function can take place on various abstraction layers as illustrated by figure 2. Ideal element behavior can be used for basic design function verification. More sophisticated models support the development and verification of analog system regarding non-ideal elemental function. Due to

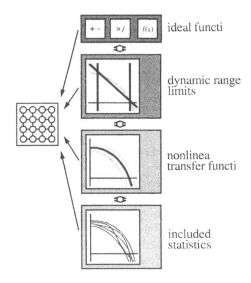

Fig. 2   Different Types of Models

this trade-off simulation performance and accuracy are exchangeable.

Nonlinear models can be realized by either analytical or numerical representation. The former approach is suitable for a wide range of practical applications but can often fit real behavior only within a limited interval of input signals.

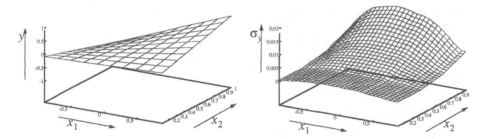

Fig. 3   Model Representation Example of a two-quadrant multiplier: multiplication result (left) and standard deviation of the multiplication error (right)

Another possibility is the usage of lookup tables. The sample values are obtained by means of standard circuits simulators. Values between samples can be interpolated using linear or cubic spline algorithm. The latter one provides smooth models with better quality at the expense of computation time.

As mentioned before, the element behavior can differ between cells due to device parameter variation. Additionally random noise can be superimposed to the signals. These effects have to be included in the analysis of parallel analog systems and thus, have to be modeled appropriately.

In order to do so, the circuit implementation of a processing element is analyzed and the influence of device tolerances is evaluated. Thus, a behavioral model in terms of an average

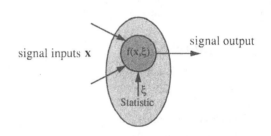

Fig. 4   Simulation Node with Statistic Variables

transfer function and its standard deviation. As example figure 3 illustrates the transfer function and the signal dependent standard deviation of a real multiplier circuit's model [4].

For the statistical modeling random variables are included in the functional description of the model (figure 4). They represent additional input signals with defined statistical behavior in terms of distribution, parameters and timing behavior. Fast changing statistic variables can represent different types of noise and static variables model device mismatch.

4   Simulator Realization

An object oriented approach is the base of the new simulation environment. As mentioned before, the elements have to be represented as nodes in order to apply a relaxation method. A node is realized as object with standardized interface. This interface allows the interconnection of nodes and the calculation of simulation steps. The different types of nodes are characterized by their individual transfer functions which contain detailed signal processing instructions and random variables if required.

Prototypes for hierarchies and regular fields support especially highly parallel systems in this simulation environment. With this tool the systems network is represented as object oriented software code that is compiled monolithically within the simulation environment. For realization ANSI C++ programming language was selected because this language already contains advantages for development of portable and reusable code. By compilation of the C++ source code a compact binary code is created, which is executed very fast, in contrast to script-controlled interpreting simulation environments.

netlist based simulation source code

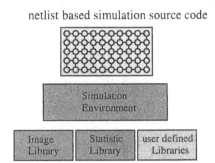

Fig. 5   Simulation System Structure

In addition to the achieved simulation speed by compilation a wide range of other C/C++ libraries can be used to extend the functionality of the simulation system. Image libraries allow direct import and export of images where pixel values contain stimuli or calculation results. Statistical libraries can help to generate stimuli and enable the statistical analysis to be done within the simulation environment. Other user defined analysis libraries can be added easily as well. These possibilities of user defined data processing within the simulation environment facilitate a fast and efficient system analysis (figure 5).

5   Design flow

The proposed design and simulation process of highly parallel analog systems benefits from the combination of conventional design tools and the new developed simulator. The circuit implementation are designed with the help of standard design tools as Cadence. But also network topologies or functional blocks can be designed and simulated with these tools. Obviously, the cell-libraries have to exist in both simulator environments. The netlist of network or block description generated by the graphical design tool can be imported directly into the dedicated

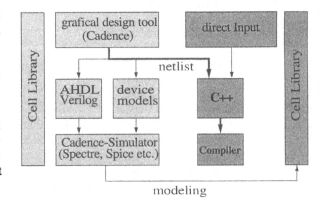

Fig. 6   Design Flow

simulator. They are translated to C++ source code and compiled with the basic C++ simulation libraries. Additionally, the generation of code from abstract mathematics is possible as well.

So, new simulation environment can be used with either bottom-up or top-down design methods. It is no replacement for conventional simulation tools but a extension of the design environment to enable a efficient simulation of highly parallel systems. Figure 6 illustrates the design flow.

The cell libraries containing the behavioral models of the function elements can also be transferred to standard simulation tool. Thus it is possible to simulate the same topology with the same behavioral model using both tools. This has be done in the following example.

6   Results

The advantages and potentials of the new simulation system shall be proved with simulation of an analog image processing array comprising of 16k processing elements. Each element consists of three multipliers, four adders and a non-linear function. This processing array is a part of a CMOS image sensor N128 [5] with preprocessing.

A behavioral model of the processing element was derived and implemented in both simulation systems Cadence and the new simulator. With the latter one the simulation of a 35x17 element array was accomplished in a fraction of time as outlines in table 1. Larger arrays could not be simulated with the Cadence tool due to limited resources. Additionally, the input and output of images with this tool is rather complicated.

	Spectre	new Sim.	
size (in elements)	35 x 17	35 x 17	128 x 128
memory usage	1,2 GB	51 kB	2 MB
nodes equations	10219 35804	- 595	- 16384
calculation time	3928 s	< 1 s	10 s

Table 1:Comparison of N128 Simulation

The new simulation tool does not have these limitation and is able to simulate 128x128 element array as implemented in the N128. The results are shown in table 1 for comparison.

7   Current Development

Current developments are aimed to a wide range of optimization. This includes improvements of actual software design but also optimization of methodical approaches. The relaxation method itself can be improved i.e. by usage of the Gauß-Seidel method which offers better convergence. A substantial acceleration of the simulation process is possible by usage of graph theory for network optimization. This method optimizes the network structure itself and thus, can also result in more efficient hardware design. But to evaluate the possibilities of this approach more scientific research is required.

Furthermore the support of different types of hardware description languages and netlists needs to be extended. This offers an easy usage of the simulation system in different design environments. Improvements in the area of user accessibility and controllability are also planned.

8   Conclusion

In this paper the difficulty of simulating highly parallel analog systems with standard circuit simulators was described. We proposed a new simulation system. This system is able to process even very large networks fast and with affordable accuracy. The possibility of easy extension is another advantage. Due to standardized data exchange it is usable with many available standard simulation tools.

Acknowledgment

This work was supported by the Sonderforschungsbereich 358, Teilprojekt A7 of the Deutsche Forschungsgemeinschaft.

Literature

[1] R. Saleh, S. Jou, A.R. Newton: Mixed-Mode Simulation and Analog Multilevel Simulation, Kluver Academic Publisher, 1994

[2] R. S. Varga: Matrix Iterative Analysis, Prentice-Hall, 1962

[3] J. M. Ortega, W. C. Rheinbolt: Iterative Solution of Nonlinear Equations in Several Variables, Academic Press, 1970

[4] A. Graupner and R. Schüffny: An Ultra-Low-Power Switched-Current 2-Quadrant Multiplier, 2nd Electronic Circuits and Systems Conference ECS'99, Bratislava, September 6-8, 1999.

[5] J.-U. Schlüßler, I. Koren, J. Werner, J. Dohndorf, U. Ramacher, A. Krönig: Image Sensor with Analog Neural Preprocessing, MicroNeuro'97, Seiten 209-216, 1997.

SYSTEM SIMULATION AND EVALUATION

EVALUATION

Methods and Tools

Object Oriented System Simulation of Large Heterogeneous Communication Systems

Uwe Hatnik, Jürgen Haufe, Peter Schwarz
Fraunhofer Institut für Integrierte Schaltungen, Germany
email: hatnik@eas.iis.fhg.de

Abstract

Communication systems consist of many soft- and hardware components with a wide range of parameters which affect mainly the provided quality of service. One of the main challenges for configuration and structuring such a heterogeneous system is to guarantee the specified quality of service with a minimum of costs. In this paper, we introduce a simulation based approach which helps the designer to determine the best fitting parameter values. Our approach combines prototyping and simulation in a common environment.

1 Introduction

Modern communication systems are very complex heterogeneous systems realizing world-wide video and audio communication and using different networks and protocols with a specified quality of service. Such communication systems consist of servers and clients. Especially clients are very different user devices, from powerful personal computers to small cellular phones. A client can communicate with other clients and servers, using services like live video conferences or it can store and can demand video and audio records (see also Figure 1). Clients and servers are connected by common local and wide area networks.

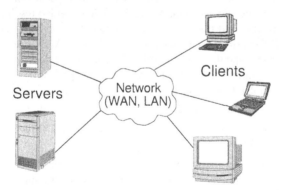

Fig. 1: Basic elements of a communication system.

One of the main challenges for configuration and structuring of such a heterogeneous system is to guarantee the specified quality of service with a minimum of costs. The designer may meet the challenge by using his practical knowledge or by building up prototypes or by utilising formal methods such as performance analysis and simulation.

185

R. Merker and W. Schwarz (eds.), System Design Automation, 185-194.
© 2001 *Kluwer Academic Publishers.*

In this contribution, we introduce a simulation based analysis approach which combines the fore-mentioned analysing methods. In our approach both simulation models and real hardware and real software prototypes can be executed in a common environment. Results of the application of formal methods may be integrated into the simulation models, e.g. distribution functions, profiling results as well as measured values. The approach was driven by our experience that only a mix of different analysis methods which complement one another may bridge the analysis gap of such huge heterogeneous systems.

The text is organized as follows. Section 2 details the analysis requirements of the system we focus on. Section 3 gives an overview of our modelling approach. Implementation aspects are described in section 4.

2 Requirements in communication system design analysis

Clients and servers of a communication system consist of software and hardware components like real-time and non real-time operation systems [3], conference system applications [1], data bases, network processors [5], data processing units, coprocessing hardware[2] [4], RAM, and persistent memory. Some parts of the system are already implemented, others exist only as abstract or detailed models. The system designer has to find an optimal configuration in accordance with the requested parameter. He has also to determine all important parameters of a given system configuration to check the system reserve. The main target is then the determination of an optimal system load with minimal soft- and hardware costs as well as low power dissipation. To figure out the system parameters the designer can measure, analyse and simulate the whole system or parts of it. In practice a complete assessment of the whole system will only be possible with a mixture of these methods. Therefore we propose a procedure that combines all three methods, e.g measuring the already implemented components and analysing or simulating the non-existing parts in such way, that the approaches complement each other. The main objectives of such an simulation environment are:

Parameter determination: A lot of parameters influence the system behaviour. One goal of the system simulation is to find optimal parameter values for a special configuration. Some parameters are specified by the service demand, for example the video resolution, the number of colours, the net bandwidth, and the used network protocol. Other parameters depend on the computer used, like CPU performance, memory size and so on. Additionally there are software parameters like buffer size and the used algorithm for data processing. There is a large amount of parameters and the optimal configuration is very system specific. Therefore the parameters can not determined completely analytically.

Performance analysis: Since optimal system parameters can hardly be determined only analytically, simulation is also important for examining system performance depending on the hard- and software parameters.

Configuration analysis: The configuration of a server or client depends on the demanded service and the client system. For example a specific data compressing algorithm is used depending on system parameters of the client like CPU performance and memory size. There are a lot of possible hard- and software combinations. It would be useful to determine what combination is suitable for a special service and configuration.

Relationships between the components: There is a more or less tight correlation between the components of a configuration. For that reason, the system has to be treated as a whole. For example, swapping out parts of the software to hardware would decrease the load of the CPU, but on the other hand, this would lead to a higher load of the system bus since of the gaining effort for communication. The simulation of the whole system allows detailed analysis of the mutual dependencies and makes it possible to separate between important and non relevant correlations.

Data transport mechanism: Data packets are exchanged between the system components. The size, organization and management of these data packages have a great influence on the system load. The simulation will help to find out and test suitable data representations and their dependence of the component granularity.

Framework: For the practical development the simulation framework will be useful for design, implementation and test of real hard- and software components, because real components can be integrated in the simulation framework.

3 Object oriented system modeling

In a first paragraph, we introduce typical system architectures of a communication system by an example. For simulation based analysing of such system architecures, the system architecture has to be transformed into a simulator specific simulation model using various modelling languages (e.g. SDL, VHDL, Verilog, Modelica). In a second paragraph, we show methods how the system architecures can be mapped into a simulation environment. Therefore, we prefer the **object oriented modelling** approach, which keeps the system hierachy, the system structure and the system component interfaces untouched when mapping the system architecture into a simulation model. Consequently, the system architecture is found again in the model structure. In fact, the object oriented modelling reduces the modelling effort as well as the error-susceptibility of the modelling process. The modelling approach includes the incorporation of existing real hardware or real software into the system model to decrease the modelling effort or to increase the simulation performance. The methodology used for the real component incorporation is more detailed explained in [9] [10].

System architecture

Figure 2 shows typical hardware structures of real clients and servers. The client in Figure 2 a) is a common personal computer. The CPU and RAM exchange data over the PCI bridge. Additional components can be coupled over the PCI system bus which is connected with the PCI bridge and provides some PCI ports (slots) for the components. In each common PC this will be a hard disc controller (e.g. SCSI) and a graphic adapter. Moreover, a conference system client needs a video card to connect a video camera to the system and a net adapter to realize the network connection. This configuration can be extended by special hardware e.g. a network processor, FPGA based prototypes or custom hardware. The purpose of this special hardware is to support the CPU because the conference system software running on the CPU contains a lot of time consuming algorithms for video and audio data processing. This algorithm (e.g. a MPEG coder or decoder) are often very complex and can overload the system resources, e.g. the CPU performance.

Other very different client structures are possible, as shown in Figure 2 b). For example the video camera can include analog and digital data processing units and already provide digital picture data over a standard interface, e.g. USB.

There are depicted two possible server structures in Figure 2 c) and d). Their base structure is similar to the clients one, but normally a server is more powerful than a client, because it has to serve many of clients. A video adapter, like the clients use, is usually not necessary because the server stores and provides video data but does not generate them. The server software contains a database with all stored records. If a client ask for a video, the server searches it in the database as well as initializes and controls the data transmission if it can provide the requested service. The network which connects clients and servers may have very different topologies. For example it can be a local area network but also a wide area network like the internet. The transmission way can include different technologies like cable, fiber and radio as well as different protocols.

Model architecture

The system consists of many heterogeneous components, like shown in Figure 2. Our object oriented model approach use this system structure to build a model structure, like illustrated in Figure 3. The component models can consist of a hierarchical structure of submodels and be realized using different abstraction levels.

As briefly described in the paragraph above, there is no unique architecture and no unique data protocol format for a today's communication systems. The system components are very heterogeneous. Therefore, a wide range of modelling means has to be considered to map the system into an simulation environment. Furthermore, depending on the simulation goal, the model abstraction may be high or low. Figure 3 shows as an example some typical system views. The model abstraction starts with abstract queuing models describing statistical data rates and delays and ends with a detailed descriptions of hardware blocks using hardware description languages such as VHDL or software components using programming languages. For system analysis it is important to handle all this abstraction layers within a common environment.

The most abstract model approach is to consider all or some of the clients and/or servers as traffic generators and sinks, like shown in Figure 3 a) and d). For example a media server which

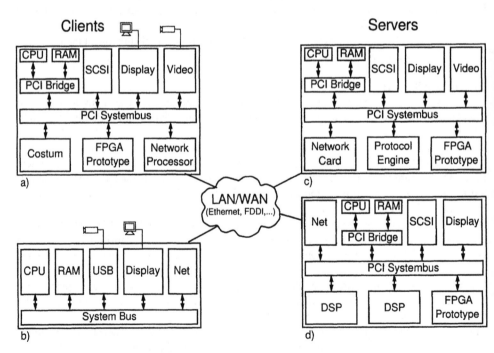

Fig. 2: Client and Server Structure

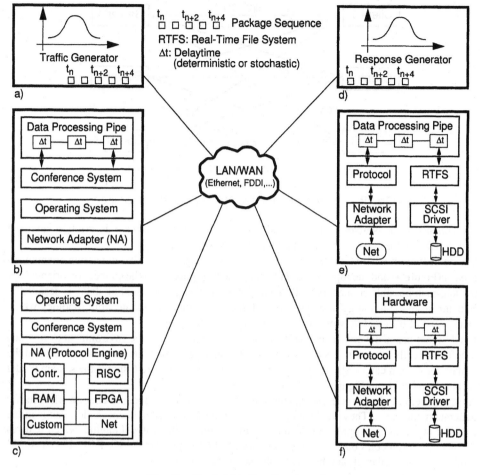

Fig. 3: Model Abstraction Levels

sends data to a client will put data packages in the network and the client will receive it. This process can be described with some parameters like package size and package distribution function. This parameters can be determined analytically or by measuring of a already existing system. Another abstraction level is shown in part b) and e) of Figure 3. In this case the model is built with soft- and hardware components. The exact model structure depends on the client or server configuration as well as on the specific simulation goal. Besides some control components this model approach includes a data processing pipe. This pipe contains all components passed by the data stream. Therefore this model type is e.g. especially suitable to observe the data flow through a data processing pipe and to determine its parameters or to test various pipe configurations. A pipe component can be an abstract model that handle abstract data or also real software.

A more detailed model approach is shown in Figure 3 c) and f). Here, one or more components are modelled very exactly. In the case of clients it is a Network Engine (c) with its components and in the case of severs it is a component from the data processing pipe (f). For example the

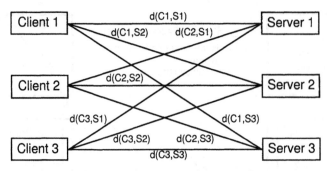

d(Cn, Sm) == link delay between Client n and Server m

Fig. 4: Connection oriented Network Model Approach

components of the Network Engine could be modelled with a hardware description language (e.g. VHDL or Verilog) or implemented with real hardware.

For the network modelling there exists two basic model approaches: the connection oriented approach (Figure 4) and the topology oriented approach (Figure 5). In the connection oriented approach, each possible connection within a network and the appropriate link delay is described. In contrast, in the topology oriented approach the model includes the network topology which can consist of different network types and their parameters.

If a designer has to develop a new hardware component he can use the environment to simulate it before it really exists. In this case the new component will be in the focus of interest and the remaining system can be abstracted to a data source or sink. Figure 6 shows the system model scenario derived from Figure 3.

Abstract clients and servers as well as the conference system of the focused client work as traffic generators and sinks. They send data packages to the more detailed client model, that is in the focus of interest, e.g. a „protocol engine". In this example the protocol engine, which realizes the network protocol, consists of six components: PCI controller, RISC, DSP, FPGA, RAM module, network adapter, and bus which connects the components.

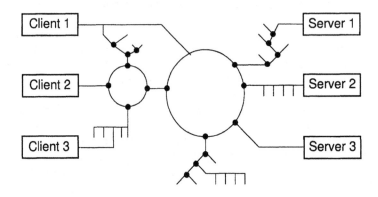

Fig. 5: Topology oriented Network Model Approach

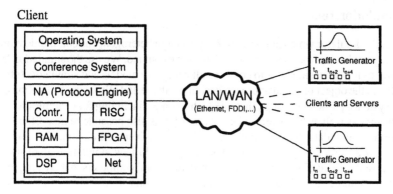

Fig. 6: Example Scenario

4 Object oriented system simulation

As shown above, a communication system contains many heterogeneous components, e.g. the network, client and server hardware components and their software. The simulation of each single component or of component groups is state of the art. A simulator which allows to simulate the whole system is currently not known. Therefore we use a **object oriented simulation** approach to develop a large heterogeneous simulation framework.

4.1 Simulation Approach

In the last years, the object oriented simulation approach has been established and also our approach follows this idea [7]. Object oriented simulation means that a system is split into subsystems, which are simulated autonomously [17]. Such a subsystem is named "object" and may contain other objects, so that an object hierarchy may be built. Figure 7 shows the object structure (a) and an object hierarchy (b). An object embeds its own simulation algorithm, which can be a small code fragment but also an extensive simulator. All implementation details are encapsulated by the object, only an interface allows data exchange and simulation control.

The advantage of that approach is its flexibility. It makes it possible to mix and exchange objects easy, whereby very different simulation algorithms can be combined. Furthermore the simulation algorithms can be optimally adapted to the models.

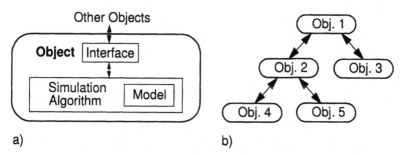

Fig. 7: Basic Object Structure (a) and hierarchical composition (b)

4.2   Simulation tools

Because a single simulator which allows to simulate a large heterogeneous communication system is currently not known, various simulators have to be coupled. We use the simulation framework **Ptolemy** as top-level simulator to combine all necessary models and simulators (objects). Ptolemy follows the object oriented approach described above and allows the simulation of heterogeneous systems, particularly those that mix technologies, including for example analog and digital electronics as well as hardware and software [6]. It is also possible to mix different simulation algorithms like data-driven simulation and discrete-event simulation. The framework provides C++ class structures that can be expanded to build new simulation algorithms and models. It is possible to distribute the simulation task to several computers that are coupled over a network. For example each client and server of the communication system can be simulated on one separate computer.

Theoretically, it is possible to develop models for each system component and simulate all within the Ptolemy framework. However this is not the goal of our work, because some components already exist and other will be developed with specific design tools.

Because some important models are only available in VHDL, we use a VHDL simulator. This permits the simulation of very large models on hardware level. Since VHDL is a very popular and flexible hardware description language, we can fall back on many own and foreign models and model libraries.

The simulation of complex networks requires a suitable network simulator and network models. We use the simulator NS (version 2), because it is very flexible and supports complex network scenarios with many transport protocols, routing mechanisms, application models etc. NS is an object oriented, discrete event driven network simulator developed at UC Berkley written in C++ and OTcl [15]. It is very useful for simulation of local and wide area networks including mobile and satellite networks.

A network scenario consists of three levels. The first level defines the network type and topology (e.g. a LAN with CSMA/CD MAC protocol), using nodes and links which connect the nodes. The second level defines the used transport protocols (e.g. UDP or TCP). Finally the application-level defines applications (e.g. telnet and ftp), which use the transport protocols.

Main simulation goal is to obtain information about the availability, workload, efficiency, and failure behaviour of the simulated scenario, dependent on a lot of parameters (link bandwidth, transport delay, protocols etc.).

The fast simulation of complex processors (e.g. DSP or RISC) requires instruction set simulators (ISS) [14]. An ISS simulates the program running of a processor. The ISS reads commands and data from a memory model, performs the data processing, and writes the results back to the memory. Neither the memory nor the processor is normally modelled on hardware level, but more abstract data processing models are used. This allows a high simulation performance, because the simulation of the exact hardware behaviour is not necessary. The abstract model level requires an adaptation if an ISS has to be combined with more detailed models, e.g. a memory model designed with VHDL.

Figure 8 shows one possible simulator structure. It is a combination of the simulation framework Ptolemy, a VHDL simulator (VSS), a network simulator (NS v2), and an instruction set simulator. Furthermore a real hardware component is included. In this example we assume that the PCI controller, network adapter, conference system, and other clients and server models are available as C code. The FPGA and RAM modules are described in VHDL and for the RISC and DSP are special instruction set simulators available. The concept is not limited to the listed simulators, so also other tools (e.g. a Verilog simulator) can be included.

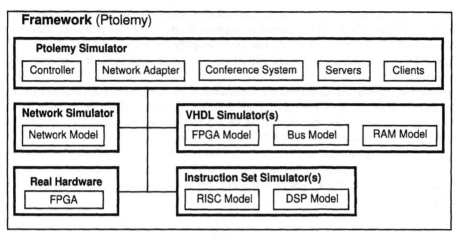

Fig. 8:   Example Simulator Structure

4.3   Simulator coupling

As seen in Figure 8, more than one simulator is necessary to simulate the whole system. There-fore, different simulators must be coupled. For simulator coupling different techniques can be used. To simplify this task, we looked for a standardized interface to couple simulators. Com-mercially, the **backplane** approach is widely spread [16]. The backplane is a separate and sim-ulator independent software to manage the communication between the simulators. As a disadvantage, the backplane produces a high communication overhead so that the relevancy of that technique is decreasing. To overcome the overhead while keeping the independency, ven-dor and simulator independent standard interfaces between complex simulation models and simulators are being developed such as the **Open Model Interface** OMI [11] [13].

The OMI, since 1999 a IEEE standard (P1499), was developed to couple simulators with com-plex digital models. It is an open standard interface, which allows interoperability between an application (e.g. a simulator) and functional models, which are presented in a binary form. The models can be developed in a variety of languages, for example VHDL, Verilog HDL, and C. To generate a binary form, a suitable compiler is necessary. Because models are supplied in a binary form, a simple but effective IP (intellectual property) protection is possible. Furthermore, a single model library may contain models of different model providers. These multiple models may be used concurrently during one session.

A separate software component exists between application and models, the OMI model manag-er. The model manager connects the application with the models and provides built-in or exter-nal models. If the application wants to use a model, the model manager creates and manages a unique customized instance, derived from the demanded model. Figure 9 shows the resulting structure. Only the interface between application and model manager is defined by the OMI specification. The connection between model manager and models as well as instances is unde-fined. This allows a very high flexibility as well as easy model generation and reuse. The model manager mainly provides routines to realize the functionality of the models and to deliver model information. Furthermore the model manager may implement additional functions, for example version and licence management. A model query mechanism may be used by the application to get model information, e.g. ports, parameters, supported data types, viewports etc. Viewports are model-internal ports, visible from the outside to simplify verification and test. They have to be provided by the model developer and allow to control the internal visibility with respect to IP protection.

The interaction between the model and the application is not fixed but can be adapted dynamically during runtime. A so-called callback mechanism allows to register and remove model function calls (named callbacks), dependent on the required interaction. More information about the goals and advantages of the OMI can be found in [12].

We use the OMI approach not only to couple simulators with functional models, but also with other simulators. This is possible, because the Open Model Interface (OMI) supports the described object oriented simulation approach, so the applied simulators can be encapsulated in an object.

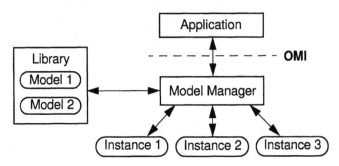

Fig. 9: OMI Basic Structure

In our simulation environment the application is the simulator Ptolemy working as "master" simulator. All other simulators are simulation objects as depicted in Figure 7. Therefore we have developed a suitable interface to couple NS with other simulators, e.g. a VHDL simulator. The interface allows to replace one or more application and protocol models with external models, which can be performed using another simulator, e.g Ptolemy or VSS.

Figure 10 shows the basic structure if Ptolemy objects use external applications via the OMI. The OMI provides the external objects to the Ptolemy models (PCI controller, network adapter, conference system as well as other clients and server). These models can use the FPGA, RAM, and Bus model that runs on an external VHDL simulator, the network model running on the network simulator, the RISC and DSP model simulated with special instruction set simulators, as well as a real FPGA. The Ptolemy models "see" only the Open Model Interface but not the external model implementation.

4.4   Integration of real hardware

In [8], we have shown the possibility to include real hardware into simulation environments via the OMI, exemplarily based on the interface board SimConnect. This board controls the data transfer between a host computer and a FPGA prototyping board. An application programming interface (API) realizes data transfer and control functions, which are executed by the SimConnect board. The SimConnect board realizes data exchange between host and FPGA and controls the FPGA prototyping board according to the called API commands. A more detailed description can be found in [9].

Object-oriented Modelling of Physical Systems with Modelica using Design Patterns

Christoph Clauß, Thomas Leitner, André Schneider, Peter Schwarz
Fraunhofer Institute for Integrated Circuits,
Design Automation Department
Zeunerstraße 38, D-01069 Dresden, Germany

Abstract

In the last years formal description languages and object-oriented design technologies became more and more important for modelling physical systems. In this paper a methodology taken from the computer science community will be presented: the Design Patterns approach by Gamma et al. [3]. Adopting to the physical systems modelling the authors propose a strategy for implementing model libraries for different physical domains. For all examples the object-oriented modelling language Modelica is used.

1 Introduction

Modelling and simulation are undoubtedly one of the key technologies in engineering. There exists a large variety of modelling and simulation environments dedicated to special physical domains. Since there is no standardized model representation, environments are often not able to interact, not even models can be exchanged.

The Modelica design group, which started as a technical committee within EUROSIM in 1996, attempted to unify different successful concepts and to introduce a common syntax and semantic. The Modelica language [1], [2] is intended to become a neutral exchange format for model representation, but it may be simulated by commercially available simulators (e.g. Dymola [9]). It shall be applicable to multiple modelling formalisms (ordinary differential equations, differential-algebraic equations, finite state automata, Petri nets etc.) and engineering domains (electrical circuits, multi-body systems, drive trains, hydraulics, chemical systems etc.).

In some general-purpose programming languages object-orientation has become a powerful concept which is widely accepted [13]. The same concepts could be applied successfully for modelling physical and technical systems [5], [7], [8], [10]. It improves the developer's productivity by supporting reuse and extensibility, and it provides a high level of abstraction e.g. in the modelling of complex systems. Therefore, object-orientation was decided to be the Modelica language principle. Modelica offers the class concept as its central building block of modularization. There are powerful relationships between classes for the modelling of a complex system and the system itself. Firstly, inheritance allows the derivation of similar models from base models. Existing models can be extended. Secondly, composition allows to built complex models from existing (simpler) models. The structure of such composed models can often perfectly reflect the structure of the real system.

R. Merker and W. Schwarz (eds.), System Design Automation, 195-208.
© 2001 *Kluwer Academic Publishers.*

Starting with the development of the electrical part of the Modelica standard library the authors have the possibility to verify the language's pretension. Object-oriented modelling methods are applied to electrical components which are the base for model derivation done by users. A useful library conception has to be found out. First results of the evaluation of the object-oriented method are presented. They are based on the modelling of several time-dependent sources and of semiconductor devices.

Today object-oriented programming languages like Smalltalk, C++, and Java are widely used and well established. A lot of methods for developing software systems using the object-oriented approaches are known and are discussed in the literature. One of the most interesting approach was introduced by Gamma et al. in 1995: Design Patterns [3].

One goal of the paper is to show how the object-oriented methodology and design patterns can be used also for the modelling of physical systems. With Modelica as an object-oriented modelling language it is possible to use these approaches for a unified implementation of model libraries for different physical domains. First of all a brief introduction in the main ideas of the methodology described by Gamma et al. [3] is given. In chapter 3 some fundamental concepts for modelling electrical and non-electrical systems will be explained. In chapter 4 the modelling methodology will be introduced step by step for simple and in chapter 5 for more complex examples.

2 Object-oriented Methods and Design Patterns

Based on object-oriented features like encapsulation, inheritance, and overriding the Design Pattern approach includes the following key properties:
- separation between *interface* and *implementation* of a model (this can be compared with *entity* and *architecture* in VHDL and VHDL-AMS)
- separation between a general purpose implementation (*default model*) and implementations for special cases via *adapter models*
- separation between certain properties (e.g. structural and algorithmic properties) which can be captured by different components or objects

The key point is to write most of the model code as reusable templates and then to adapt the templates for a special purpose.

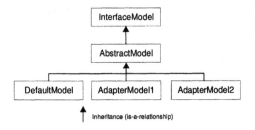

Figure 1: Inheritance hierarchy based on interface.

In **Fig. 1** the general inheritance hierarchy for a model is shown. The `InterfaceModel` only describes the connectors (pins, ports etc.), and parameters which can be seen from outside a model. The `InterfaceModel` should not presume a particular implementation. It is very similar to entities in VHDL-AMS [4]. In Modelica such an interface model can be described as a *partial model*:

```
// InterfaceModel
partial model TwoPinInterface
    Pin p;
    Pin n;
end TwoPin;
```

The Pin means here an electrical pin with voltage and current. In Modelica such pins are modelled as *connectors*. For connectors the KIRCHHOFFs laws are valid.

```
// KIRCHHOFF pin
connector Pin
    Voltage v;
    flow Current i;
end Pin;
```

The `AbstractModel` is not urgently needed but useful for many cases. This model implements general purpose features which can be used in all inherited models. For the `TwoPinInterface` example an abstract (partial) model `TwoPin` can be implemented as follows:

```
// AbstractModel
partial model TwoPin
    extends TwoPinInterface;
        Voltage v;
        Current i;
    equation
        v = p.v - n.v;
        0 = p.i + n.i;
        i = p.i;
end TwoPin;
```

In a third step a modeller can define his ready-to-use model. For model libraries it makes sense to define a model with default behaviour and in addition some adaptor models for certain cases. Now a user can take one of these models and use it directly or he can build his own model via inheritance and/or overriding mechanisms. So in Modelica a default implementation for the model `Resistor` can extend `TwoPin` as follows:

```
// DefaultModel
model Resistor
    extends TwoPin;
        parameter Resistance R;
    equation
        R * i = v;
end Resistor;
```

For models with more complex behaviour it is useful to describe certain components of behaviour in different parts. While the strategy described above (see **Fig. 1**) is purely based on *class inheritance*, the Design Pattern approach also supports the (more traditional) composition of components. Both, *class inheritance* and *object composition*, are the most common techniques for reusing functionality in object-oriented systems. Class inheritance is often referred to as

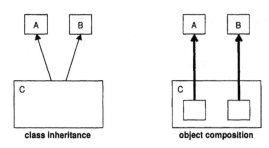

Figure 2: Class inheritance versus object composition.

white-box reuse which means, the internal structure of parent class is visible to subclasses. Object composition is an alternative to class inheritance. This style of reuse is called *black-box reuse*, because no internal details of objects are visible. Here, new functionality is obtained by assembling or composing objects to get more complex functionality. But the key point is that object composition requires that the objects being composed have well-defined interfaces.

Fig. 2 illustrates the two concepts of class inheritance and object composition. A user can build a model C using multiple inheritance from model A and model B; or he can compose model C using model A and model B as subcomponents.

Fig. 3 shows an example where the description of the algorithmic behaviour is separated from the rest of a model. The separated submodel of course can have its own class hierarchy with inheritance relationships. So the object composition can be done at different levels of abstraction.

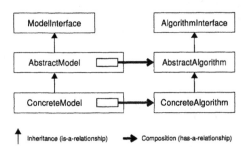

Figure 3: Separation of algorithmic model properties.

In Modelica the object composition can be described using a simple *component instantiation statement* or using the more powerful *replaceable/redeclare* keywords. The following source code illustrates the possibilities:

```
model SimpleCircuit
  Resistor R1 (R=100);
  Capacitor C1 (C=1e-6);
  // ...
equation
  connect (R1.n, C1.p);
  // ...
end SimpleCircuit;
```

```
partial model AbstractCircuit
  replaceable MOSFET T1;
  Resistor R1 (R=470);
  // ...
equation
  connect (T1.G, R1.p);
  // ...
end AbstractCircuit;

model ConcreteCircuit
  extends AbstractCircuit
  (redeclare MOSFET_Level1 T1);
  // ...
end ConcreteCircuit;
```

In this example R1 and C1 are *instantiated*, but the transistor T1 is *replaced* by MOSFET_Level1.

In addition to the design methodology Gamma et al. [3] introduce a set of well-engineered patterns that a developer can directly apply to his applications. **Table** 1 gives an overview of all patterns classified by its purpose with respect to the application.

Tabelle 1: Design Patterns.

Creational	Structural	Behavioral
Factory Method	Adapter	Interpreter
Abstract Factory	Bridge	Template Method
Builder	Composite	Chain of Responsibility
Prototype	Decorator	Command
Singleton	Facade	Iterator
	Flyweight	Mediator
	Proxy	Memento
		Observer
		State
		Strategy
		Visitor

Although the design patterns were primarily intended to use in software engineering, many of the patterns can be very useful also for modelling physical systems. In this paper the following patterns are used:

- Adapter: to adapt a general purpose model to different physical domains (e.g. signal sources: VoltageSource, CurrentSource, ...)
- Strategy: to encapsulate different algorithms for calculating a time-dependent signal (see signal sources: Function)
- Abstract Factory: to provide a unique interface for creating/using different parameter sets and equations for SPICE semiconductor models

A modeller can of course define additional patterns. The examples described in this paper are only intended to demonstrate the methodology.

3   Fundamentals of Physical Systems Modelling

Basically, the modelling process can be divided into two fundamental steps (**Fig. 4**). The second step is mostly well supported by the simulator's input processing programs. A goal of the physical modelling approach is: as much as possible modelling steps are closely related to the design process of technical systems and to the intuitive procedure of the design engineer. Essential parts of the physical modelling step are:

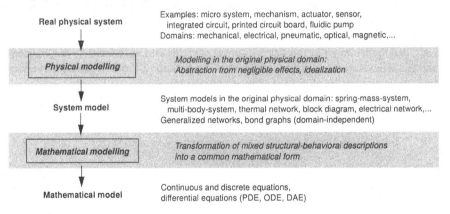

Figure 4:   Steps in modelling physical systems.

Definition of the levels of abstraction: Decisions are necessary about the first and second order effects which have to be modelled. Otherwise the accuracy has to be determined. Often some parts of the model are described very exactly using DAE's, other parts may be modelled with a more abstract description (system level).

Choice of the description method: Different description methods exist:
- behavioral description: equation-oriented, similar to the textual notation of an ordinary differential equation (ODE),
- structural description: the system model is hierarchically composed of subsystems and basic elements (the so-called primitives) available with the simulators.

Mostly, system models with spatially lumped elements appear as implicit nonlinear differential-algebraic equations (DAE). Only in special cases their formulation as explicit state equations is possible. Well-known description means are mathematical descriptions (for tools like Mathematica or Maple), hardware description languages, block diagrams, or networks (e.g. for SPICE).

Definition of the interface : The ports to the surrounding have to be divided into conservative ports (that meet KIRCHHOFFs laws) or nonconservative ports.

Definition of the signal properties: Investigations are necessary on the time variable, and on the signal values which can be discrete or continuous.

The **multiport modelling approach** is a very fundamental, systematic, and powerful method. The complete system is decomposed into subsystems (multiports). Signals are separated into *external* signals between the subsystems, and *internal* signals within the subsystems. The external signals of a subsystem form its *terminal signals*. The behaviour of subsystems depends only on their terminal signals (and some internal signals) and their initial conditions, too. Each subsystem can be substituted by other subsystems. The behaviour of a subsystem can be expressed

- implicitly by the connection of other subsystems: hierarchical structural refinement,
- explicitly by a set of equations (e.g. nonlinear differential-algebraic equations): „behavioural modelling" in a strict sense,
- or a combination of both of them if the simulator has language constructs to formulate mixed structural-behavioral descriptions.

The key problem in setting up the system equations is the description of the terminal behaviour of the *subsystems*. It has to be investigated according to physical laws. An external view of a multiport is shown in **Fig. 5**. The signals are vector-valued functions of time t. The terminal signals are divided into:

- v_1, i_2, a_{in}, d_{in}: independently chooseable difference, flow, and non-conservative signals
- v_2, i_1, a_{out}, d_{out}: dependent difference, flow, and non-conservative signals

In many cases it is impossible to state the terminal behaviour only based on terminal signals. Subsystems may then have internal states s. This multiport modelling approach is closely related to *object-oriented modelling* in the sense of the construction of strictly *hierarchical, modular models*.

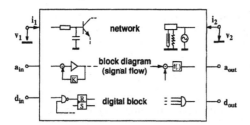

Figure 5: Different kinds of terminal signals of a multiport

4 Modelling of signal sources

For modelling electrical and non-electrical systems various time-dependent signal sources are needed. In the context of implementing a Modelica standard library for electrical components [6] the authors have designed a flexible model hierarchy for signal sources which can easily be adapted to different physical domains. In the following the most important classes (models) of the hierarchy will be explained and the a vantages of the Design Pattern approach will be shown. **Fig. 6** gives a first impression: classes will be introduced from the inner to outer block.

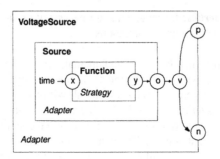

Figure 6: Components of VoltageSource.

In the core of each signal source a function block realizes the signal behaviour dependent from time *t* or another independent variable *x*.

$$y = f(x)$$

First of all an interface for this block will be formulated in Modelica as follows:

```
partial model Function
   input Real x;
   output Real y;
end Function;
```

Now this interface can be used for certain implementations. Because of the simplicity of Function an abstract or default implementation (as introduced in chapter 2) is not necessary. As an example, the implementation for a *Sine* function is given:

```
model Sine extends Function
   parameter Real a=1;
   parameter Real b=1;
   parameter Real c=0;
equation
   y = a * sin(b * x + c);
end Sine;
```

A second example implements the non-continuous *Step* function:

```
model Step extends Function
   parameter Real x_step=0;
   parameter Real y_pre=0;
   parameter Real y_post=1;
equation
   y = if (x < x_step) then y_pre
          else y_post;
end Step;
```

In a function library all interesting signal behaviours can be modelled as pure mathematical functions.

For using the function block in block diagrams of schematic editors or netlists it is necessary to convert the plain type Real to an Modelica connector:

```
connector RealOutputSignal
   output Real s;
end RealOutputSignal;
```

With this connector a new interface is modelled:

```
partial model SingleOutput
   RealOutputSignal o; // output connector
end SingleOutput;
```

Because time-dependent signal sources are the most important sources for modelling physical systems, now the abstract model Source will be introduced via *class inheritance* (extends). This model uses the Function interface as replaceable subcomponent (see *object composition*, chapter 2) and assigns the current time (time is a predefined variable in Modelica which gets the current time) to the input x of Function.

```
partial model Source
   extends SingleOutput; // interface
   replaceable Function f; // interface
equation
   f.x = time;
   f.y = o.s; // output
end Source;
```

A complete signal source can be obtained by redeclaring Function with a certain implementation (e.g. Sine):

```
model SineSource
   extends Source (redeclare Sine f)
end SineSource;
```

Finally for using the signal sources in a certain physical domain an adapter must be modelled. The goal of the adapter should be the conversion of all parameters and input/output signals to a concrete physical context. The implementation of such adapters is demonstrated for voltage source and current source. The two models are abstract implementations which extends the abstract model TwoPin (see chapter 2).

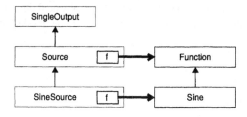

Figure 7: Relationships for the signal source.

```
model VoltageSource extends TwoPin;
   replaceable Source f;
   parameter Voltage V;
equation
   v = V* f.y;
end VoltageSource;
```

```
model CurrentSource extends TwoPin;
  replaceable Source f;
  parameter Current I;
equation
  i = I * f.y;
end CurrentSource;
```

By redeclaring (instantiating) Source with a concrete signal source (e.g. SineSource) a user can obtain a ready-to-use model.

```
model SineVoltage
  parameter Frequency freqHz = 1;
  parameter Real phase = 0;
  extends VoltageSource
    (redeclare SineSource
    f(b=2*pi*freqHz,c=phase));
end SineVoltage;

model SineCurrent
  parameter Frequency freqHz = 1;
  parameter Real phase = 0;
  extends CurrentSource
    (redeclare SineSource
    f(b=2*pi*freqHz,c=phase));
end SineCurrent;
```

Fig. 8 illustrates the relationships for the voltage source.

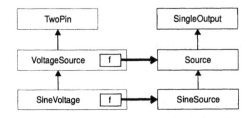

Figure 8: Relationships for the voltage source using signal sources (see **Fig. 7**).

Now it is very simple to model signal sources for other physical domains. For example, a TemperatureSource and a SineTemperature can be modelled as adapter models as follows:

```
model TemperatureSource
  extends ThermalTwoPin;
  replaceable Source f;
  parameter TemperatureC T;
equation
  temp = T* f.y;
end TemperatureSource;

model SineTemperature
  parameter Frequency freqHz = 1;
  parameter Real phase = 0;
  extends TemperatureSource
    (redeclare SineSource
    f(b=2*pi*freqHz,c=phase));
end SineTemperature;
```

5   SPICE-like Semiconductor Modelling

In the field of semiconductor models SPICE model equations have become a quasi standard for a long time. They are the almost only used models in industry besides vendor specific models like Philips level 9 MOSFET model which are implemented similar to the SPICE' ones. Each SPICE model characterizes structure and equations of a device type like MOSFET or bipolar junction transistor. SPICE like modelling shall have to correspond to the SPICE modelling philosophy which is quite old fashioned but widely used. A clear concept for structuring is needed. Object-oriented modularization and programming (C++) could be used successfully in the implementation of a simulator-independent library of SPICE models [10], [11]. An object-oriented approach using design patterns will lead to a realizable modelling hierarchy.

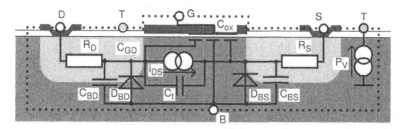

Figure 9:   Components of a MOSFET model.

Several levels of MOSFET models have been implemented to meet different needs of accuracy especially in deep submicrometer device modelling. Each model contains both standard structure components like p-n-junctions and highly specialized parts like channel charge calculation. The implementation of the latter ones consists of several hundreds or even thousands lines of source code. Thus a modularisation of the models leads to smaller but easier to handle component models (**Fig. 9**). The realization of the MOSFET channel models shown in **Fig. 10** is an example for component modelling.

Although SPICE-like simulators provide several MOSFET modelling levels differing in equations and parameter sets, all levels are combined into one MOSFET model changing its behaviour and parameter set depending on the value of a single parameter named `level`. The object-oriented approach suggested in **Fig. 10** allows the definition of a compact MOSFET model which provides basic functionality and is extended by proper component models depending on the level of modelling chosen by the user of the model.

The final user of a SPICE semiconductor model expects SPICE-like modelling philosophy which is among others characterised by the splitting of the parameter set into device specific and technology dependent parameters.

The following source code gives the principles of a Modelica implementation providing SPICE-like MOSFET models:

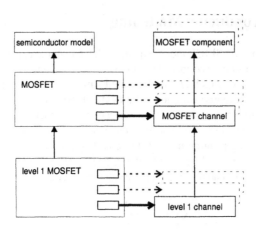

Figure 10:   MOSFET channel modelling.

```
partial model SpiceParSet
end SpiceParSet;

partial model SpiceDevice
  replaceable SpiceParSet parSet;
end SpiceDevice;

// Mosfet Parameter Set (dot models)
partial model MosParSet
  extends SpiceParSet;
  parameter int type;
  parameter int level;
parameter Modelica.SIunits.Resist RS ...;
// ... and additional parameters
end MosParSet;

partial model MosLevel1ParSet
  extends MosModel (_level=1,...);
end MosLevel1ParSet;
// MosLevel2ParSet ... Level N

// Mosfet Channel Equations
partial model MosChEq
  replaceable MosParSet parSet;
  Pin D; Pin G; Pin S; Pin B;
end MosChEq;

model MosLevel1ChEq
  extends MosChEq;
equation
  // ... level 1 channel equations
end MosLevel1ChEq;

// MosLevel2ChEq ... Level N

// MOSFET Devices
partial model Mosfet
  Pin D "Drain"; Pin G "Gate";
  Pin S "Source"; Pin B "Bulk";
parameter Modelica.SIunits.Length L=...;
parameter Modelica.SIunits.Length W=...;
// ...
```

```
replaceable MosChEq chEq(parSet = parSet);
// ... other components
equation
   connect(D,equationSet.D);
   connect(G,equationSet.G);
   connect(S,equationSet.S);
   connect(B,equationSet.B);
   // ... other components connects
end Mosfet;

model MosfetLevel1
   extends Mosfet( L=... ,W=..., ...,
   redeclare MosLevel1ParSetparSet,
   redeclare MosLevel1ChEqchEq);
end MosfetLevel1;

// MosfetLevel2 ... Level N

model ExampleCircuit
   // ...
   MosfetLevel1 tr1(...);
   // ...
end ExampleCircuit;
```

6 Summary and Outlook

The main ideas of the object-oriented modelling strategy and the theoretical fundamentals for analysing and formal description of physical systems were discussed. A methodology for implementing model libraries was introduced and illustrated with some simple examples. The most important advantages are:

- clear concept for hierarchical structuring of large model libraries
- simultaneous implementation of models for different physical domains using a unique core library with basic algorithms (phenomena in different physical domains often have the same mathematical background)
- efficient implementation because of reusing code (class inheritance, object composition)
- extensibility of model libraries at different levels of abstraction

In certain cases a disadvantage of the demonstrated methodology could be the distribution of knowledge to many parts of a library. So sometimes it could be better if you have only one monolithic model which has all the equations and modelling knowledge inside. But the authors believe that in future flexibility and extensibility of models and model libraries become more and more important. To reach this goal also other approaches and languages (like VHDL-AMS) have to be examined.

References

[1]   Elmqvist, H. et al.: Modelica - A Unified Object-Oriented Language for Physical Systems Modeling. Version 1.3, December 1999.
      http://www.Modelica.org
[2]   Elmqvist, H.; Cellier, F.; Otter, M.: Object-oriented modeling of hybrid systems. Proc. European Simulation Symposium, ESS '93, 1993.

[3]   Gamma, E.; Helm, R.; Johnson, R.: Design Patterns - elements of reusable object-oriented software. Addison-Wesley, Reading 1995.

[4]   IEEE: Standard VHDL Analog and Mixed-Signal Extensions. Technical Report IEEE 1076.1, 1997.

[5]   Kecskemethy,A.; Hiller, M.: An object-oriented approach for an effective formulation of multibody dynamics. 2nd US Natl. Congress Computational Mechanics, Washington, 1993.

[6]   Modelica Standard Library v1.3.1:
      http://www.Modelica.org

[7]   Otter, M.: Objektorientierte Modellierung mechatronischer Systeme am Beispiel geregelter Roboter. Dissertation, Bochum 1994.

[8]   Kasper, R.; Koch,W.: Object-oriented behavioral modelling of mechatronic systems. Proc. 3rd Conf. Mechatronics and Robotics, Paderborn 1995, 70-84

[9]   Dymola: http://www.Dynasim.se

[10] Leitner, Th.: Entwicklung simulatorunabhängiger Modelle für Halbleiter-Bauelemente mit objekt-orientierten Methoden. Berlin, Verlag dissertation.de, 1999.

[11] Leitner, Th.: A simulator independent semiconductor model implementation based on SPICE model equations. Analog Integrated Circuit and Signal Processing Journal, Volume 21, No. 1, October 1999, Kluwer Academic Publishers.

[12] Bergé, J.-M.; Levia, O.; Rouillard, J.: Object-Oriented Modeling. Kluwer Academic Publishers 1996.

[13] Booch, G.: Object Oriented Design with Applications. Benjamin/Cummings, Reading MA, 2nd Ed., 1994.

Simulation of Numerically Sensitive Systems by Means of Automatic Differentiation

Klaus Röbenack and Kurt J. Reinschke, TU Dresden, Institut für Regelungs- und Steuerungstheorie

Abstract

Differentiator blocks are useful in many applications. However, their use leads to the simulation of a numerically sensitive system since they use difference formulas. We suggest a new method which provides exact derivative values. Our approach is applicable to simulators based on object oriented programming languages.

1 Introduction

Block-oriented simulation packages are widely used by engineers. Typical blocks are integrators and nonlinear functions. The inputs and outputs of these blocks are function values of the associated signals at discrete points of time. Dynamic blocks (e.g. integrators and delay time blocks) store past values of the input or output signal and use them for the computation of the actual output value (see Fig. 1).

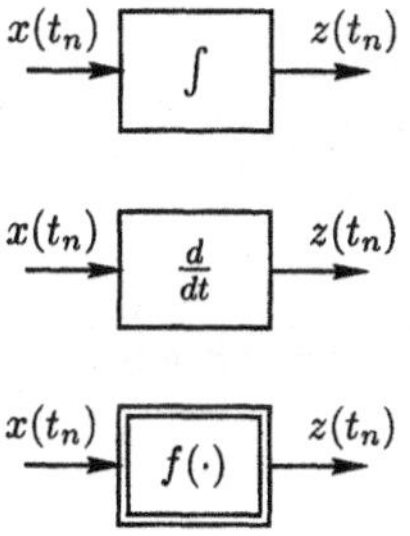

$$x(t_n) \rightarrow \boxed{\int} \rightarrow z(t_n) \qquad z(t_n) = z(t_{n-1}) + h\,x(t_n)$$

$$x(t_n) \rightarrow \boxed{\tfrac{d}{dt}} \rightarrow z(t_n) \qquad z(t_n) = \tfrac{1}{h}\left(x(t_n) - x(t_{n-1})\right)$$

$$x(t_n) \rightarrow \boxed{f(\cdot)} \rightarrow z(t_n) \qquad z(t_n) = f(x(t_n))$$

Figure 1: Some blocks and the computation of the output signals with a step size h

Modern control concepts like feedback linearization [1,2] and flatness-based control [3–6] as well as the inverse system approach in communication theory [7,8] require block devices for differentiation. Simulation programs approximate time derivatives with difference formulas like

$$\dot{x}(t_n) \approx \frac{x(t_n) - x(t_{n-1})}{t_n - t_{n-1}}.$$

This approach is very sensitive to numerical errors. It is almost impossible to compute higher order time derivatives in this manner.

R. Merker and W. Schwarz (eds.), System Design Automation, 209-220.

Example 1 We will explain the problems described above by a chaotic transmission system based on the *Rössler attractor* [9–12]

$$
\begin{aligned}
\dot{\xi}_1 &= -\xi_2 - \xi_3 \\
\dot{\xi}_2 &= \xi_1 + a\,\xi_2 \\
\dot{\xi}_3 &= b + (\xi_1 - c)\,\xi_3 - u \\
y &= \xi_2
\end{aligned}
\tag{1}
$$

with the parameters $a = 0.55$, $b = 2$, $c = 4$, the additional input u and the output y. To design the inverse system we transform the differential equation (1) into normal form

$$
\begin{aligned}
\dot{x}_1 &= x_2 \\
\dot{x}_2 &= x_3 \\
\dot{x}_3 &= f(x_1, x_2, x_3) + u \\
y &= x_1
\end{aligned}
\tag{2}
$$

with $f(x_1, x_2, x_3) = -cx_1 + (ac - 1)x_2 + (a - c)x_3 + (ax_1 - x_2)(-x_1 + ax_2 - x_3) - b$. The relative degree [1] of this system is three. Hence, the inverse system contains three differentiation blocks [7, 8]. The block diagram of the whole transmission system is shown in Fig. 2. The test input signal is $u(t) = \gamma \sin(\omega t)$ with $\gamma = \omega = 0.05$. The input signal u (transmitter) and the reconstructed input signal u_r (receiver) should coincide. The simulation based on the default settings of MATLAB/SIMULINK (**ode45** with variable step size) and the initial values $x_1 = x_2 = x_3 = 0$ gives the results shown in Fig. 3. Obviously, the signals u and u_r differ significantly from each other. □

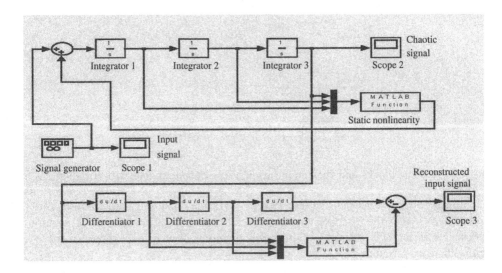

Figure 2: Chaotic transmission system

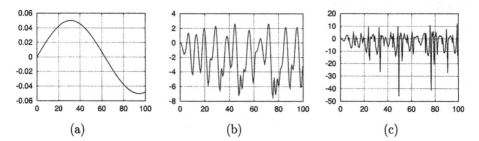

Figure 3: Simulation results: (a) input signal u, (b) output signal y, (c) reconstructed input signal u_r

2   Automatic Differentiation

In this section we sketch the concept of automatic differentiation techniques [13–18]. There are many applications where for a function F one needs not only function values but also values of the derivatives. The most obvious way to get a derivative is the symbolic differentiation with software packages such as MATHEMATICA, MAPLE and MUPAD. However, for higher order derivatives the computational effort (computation time, memory requirements) can increase drastically due to chain, product and quotient rules [19]. Furthermore, symbolic differentiation usually entails a repeated evaluation of common expressions. On the other hand, difference quotients like

$$\frac{F(x+h) - F(x)}{h} \quad \text{or} \quad \frac{F(x+h) - F(x-h)}{2h}$$

do not provide accurate values because of cancellation and truncation errors. For higher order derivatives these accuracy problems become acute. These disadvantages of symbolical and numerical differentiation can be circumvented with *automatic differentiation* [20]. Similar to the symbolic differentiation, elementary differentiation rules like sum, product and chain rule will be applied systematically. Tools for automatic differentiation will use concrete numbers instead of symbolic expressions.

There are two basic concepts for the implementation of automatic differentiation software [20]: *program transformation* with compiler-generators and *operator overloading* in case of object oriented programming languages like C++ or Fortran90. In the last case, one has simply to replace the floating point type `double` by a new class (say `Ddouble`. This new class must contain both the function value (`val`) and the value of the derivative (`der`):

```
class Ddouble
{
public:
   double val;
   double der;
};
```

Elementary functions (e.g. sin, cos, exp, log) have to be redefined for this class, for instance:

```
Ddouble sin (Ddouble x)
{
 Ddouble z;
 z.val = sin (x.val);
 z.der = cos (x.val) * x.der;
}
```

Function values can be calculated as usual. Derivative values have to be computed according to the differentiation rules. Operator overloading is very useful for binary operations (e.g. $+$, $-$, $*$ and $/$). For example, the multiplication could be defined as follows:

```
Ddouble operator* (Ddouble x, Ddouble y)
{
  Ddouble z;
  z.val = x.val*y.val;
  z.der = x.der*y.val+y.der*x.val;
  return z;
}
```

Moreover, operator overloading can be employed to compute univariate Taylor series. Consider a smooth map $\mathbf{F} : \mathbb{R}^n \to \mathbb{R}^m$ which maps a curve $\mathbf{x}$ into a curve $\mathbf{z}$. Let $\mathbf{x}$ be given by

$$\mathbf{x}(t) = \mathbf{x}_0 + \mathbf{x}_1 t + \mathbf{x}_2 t^2 + \cdots + \mathbf{x}_d t^d \tag{3}$$

with vector-valued coefficients $\mathbf{x}_0, \ldots, \mathbf{x}_d \in \mathbb{R}^n$. We will use the function $\mathbf{F}$ to map the curve $\mathbf{x}$ into a curve $\mathbf{z}$. Provided that $\mathbf{F}$ is d times continuously differentiable, we can express $\mathbf{z}$ by the Taylor expansion

$$\mathbf{z}(t) = \mathbf{z}_0 + \mathbf{z}_1 t + \mathbf{z}_2 t^2 + \cdots + \mathbf{z}_d t^d + \mathcal{O}(t^{d+1}) \tag{4}$$

with

$$\mathbf{z}_j = \frac{1}{j!} \frac{\partial^j \mathbf{z}(t)}{\partial t^j} \Big|_{t=0}$$

Each Taylor coefficient $\mathbf{z}_j \in \mathbb{R}^m$ is uniquely determined by the coefficients $\mathbf{x}_0, \ldots, \mathbf{x}_j$. In particular, we have

$$
\begin{aligned}
\mathbf{z}_0 &= \mathbf{F}(\mathbf{x}_0) \\
\mathbf{z}_1 &= \mathbf{F}'(\mathbf{x}_0)\mathbf{x}_1 \\
\mathbf{z}_2 &= \mathbf{F}'(\mathbf{x}_0)\mathbf{x}_2 + \tfrac{1}{2}\mathbf{F}''(\mathbf{x}_0)\mathbf{x}_1\mathbf{x}_1 \\
\mathbf{z}_3 &= \mathbf{F}'(\mathbf{x}_0)\mathbf{x}_3 + \mathbf{F}''(\mathbf{x}_0)\mathbf{x}_1\mathbf{x}_2 + \tfrac{1}{6}\mathbf{F}'''(\mathbf{x}_0)\mathbf{x}_1\mathbf{x}_1\mathbf{x}_1
\end{aligned}
$$

Using Automatic Differentiation, the Taylor coefficients of the output signal can be obtained without symbolic computations of the derivative tensors. To calculate the univariate Taylor series of an algorithm up to the dth coefficient one could replace the type `double` by a new class `Tdouble` that contains the Taylor coefficients:

```
#include <vector>
class Tdouble:public std::vector<double>
{
public:
  Tdouble operator+
    (const Tdouble& rhs) const;
};
```

Now, one has only to overload elementary functions and operations. Some examples of the associated calculations are listed in Table 1. Using operator overloading, the Taylor coefficients of composite functions will be computed with these modified rules. A prototype implementation could have the following source code:

```
Tdouble Tdouble::operator+
  (const Tdouble& rhs) const
{
  Tdouble result;
  if( size()== rhs.size())
  {
    result.resize( size());
    iterator resultIt= result.begin();
    const_iterator thisIt= begin();
    const_iterator rhsIt= rhs.begin();
    while( thisIt< end())
    {
      *resultIt = *thisIt + *rhsIt;
      resultIt++;
      thisIt++;
      rhsIt++;
    }
  }
  else
  { // something went wrong
  }
  return result;
};
```

These rules are already implemented in Automatic Differentiation packages such as ADOL-C [21], TADIFF [22] or MIPAD [23]. Up to now, we assumed that the map **F** is sufficiently smooth. Unfortunately, the nonlinearities of technical systems often contain non-smooth functions such as saturations, min and max operations and quantizers. Then, one could replace the derivatives by weak derivatives [24] or one could replace the Taylor series by Laurent series [20, Chapter 11].

3 Simulation based on Automatic Differentiation

Our approach to avoid problems with differentiators in a block-oriented simulation is to include time derivatives in the computation. Instead of single function values $\mathbf{x}(t_n)$

Operations	Taylor coefficients
$z = x \pm y$	$z_k = x_k \pm y_k$
$z = x \cdot y$	$z_k = \sum\limits_{i=0}^{k} x_i\, y_{k-i}$
$z = x/y$	$z_k = \dfrac{1}{y_0}\left(x_k - \sum\limits_{i=1}^{k} y_i z_{k-i} \right)$ for $k \geq 1$
$z = e^x$	$z_k = \dfrac{1}{k}\sum\limits_{i=0}^{k-1} (k-i)\, z_i\, x_{k-i}$ for $k \geq 1$

Table 1: Computation of Taylor coefficients [22]

we will also take the derivatives $\dot{x}(t_n), \ddot{x}(t_n), \ldots$ into account. Because of numerical reasons, we will not compute these derivatives themselves but instead use associated Taylor coefficients $\mathbf{x}_0, \ldots, \mathbf{x}_d$.

First, we will consider a simulation scheme which consists of integrators and static nonlinearities only. The static nonlinear functions have to be realized according to Table 1. An automatic differentiation interface for MATLAB is currently under development [25–28].

In simulation schemes, integrators are basically solvers for ordinary differential equations (ODEs). Now, we will show how to employ automatic differentiation to solve ODEs. Consider an autonomous ODE

$$\dot{\mathbf{x}}(t) = \mathbf{F}(\mathbf{x}(t)). \tag{5}$$

Simultaneously, we will interpret $\mathbf{F}$ as a map between two curves $\mathbf{x}$ and $\mathbf{z}$

$$\mathbf{z}(t) = \mathbf{F}(\mathbf{x}(t)), \tag{6}$$

where $\mathbf{x}$ and $\mathbf{z}$ are given by the Taylor expansion (3) and (4). For known Taylor coefficients $\mathbf{x}_0, \mathbf{x}_1, \ldots, \mathbf{x}_i$ of $\mathbf{x}$, the Taylor coefficients $\mathbf{z}_0, \mathbf{z}_1, \ldots, \mathbf{z}_i$ of $\mathbf{z}$ can be computed with automatic differentiation as described in Section 2. Because of the identity $\mathbf{z}(t) \equiv \mathbf{x}(t)$, we can calculate the Taylor coefficient $\mathbf{x}_{i+1}$:

$$\mathbf{x}_{i+1} = \frac{1}{i+1}\, \mathbf{z}_i \tag{7}$$

As sketched in Fig. 4, we are able to compute the series expansion of the solution of Eq. 5 for a given initial value $\mathbf{x}_0$ (see [29]). As the ODE (5) is autonomous, the algorithm delivers a Taylor expansion

$$\mathbf{x}(t) = \mathbf{x}_0 + \mathbf{x}_1(t - t_n) + \mathbf{x}_2(t - t_n)^2 + \cdots$$

for a given initial value $\mathbf{x}_0 = \mathbf{x}(t_n)$ at $t = t_n$. Now, let us analyze how Taylor expansions can be used to solve ODE (5). For example, the explicit Euler method can be written as

$$\mathbf{x}(t_{n+1}) := \mathbf{x}(t_n) + h \cdot \mathbf{F}(\mathbf{x}(t_n)) \tag{8}$$

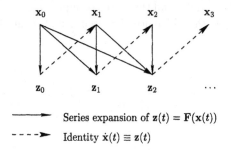

Figure 4: Signal flow for series expansion of the solution of ODE (5)

with a step size $h := t_{n+1} - t_n$. Due to the fact that $\mathbf{x}_1 = \mathbf{z}_0 = \mathbf{F}(\mathbf{x}_0)$ we can interpret (8) in terms of a Taylor series (3):

$$\mathbf{x}_0(t_{n+1}) := \mathbf{x}_0(t_n) + h \cdot \mathbf{x}_1(t_n) \tag{9}$$

In fact, there has been a growing interest in developing new integration methods based on Taylor coefficients generated by automatic differentiation [29–32].

Next, we investigate differentiator blocks. Let the input signal $\mathbf{x}$ and the output signal $\mathbf{z}$ of such a block be given by the Taylor coefficients $\mathbf{x}_0, \ldots, \mathbf{x}_d$ and $\mathbf{z}_0, \ldots, \mathbf{z}_d$. Because of $\dot{\mathbf{x}}(t) \equiv \mathbf{z}(t)$ we can apply Eq. (7) for $i = 0, \ldots, d-1$, i.e.,

$$\mathbf{z}_i = (i+1)\, \mathbf{x}_{i+1}.$$

If the number d of Taylor coefficients is large enough we should set $\mathbf{z}_d := 0$. This means that the numerical errors associated with differencing do not occur. The whole simulation scheme that combines ODE solvers and differentiator blocks is illustrated in Fig. 5.

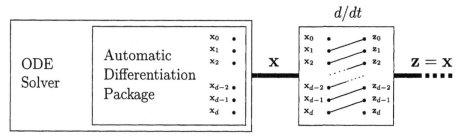

Figure 5: Simulation scheme

Example 2 Here, we will apply the results obtained in the previous section to our Example 1. As an automatic differentiation tool we used ADOL-C 1.8 with the GNU C++ compiler EGCS 2.91. The ODE (2) was solved with Euler's method (see Eq. (9)) with a fixed step size $h = 10^{-3}$. The simulation results are shown in Fig. 6. In contrast to Example 1, the input signal u and the retrieved input signal u_r do not differ visually from each other. In fact, the differences between u and u_r are less than 2.5×10^{-14} (see Fig. 7). $\qquad\qquad\qquad\qquad\qquad\qquad\qquad\qquad\qquad\qquad\qquad\qquad\qquad$ $\square$

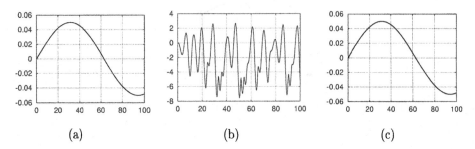

$$\text{(a)}\qquad\qquad\qquad\qquad\text{(b)}\qquad\qquad\qquad\qquad\text{(c)}$$

Figure 6: Results with the simulation scheme described in Section 3: (a) input signal u, (b) output signal y, (c) reconstructed input signal u_r

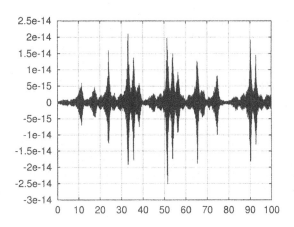

Figure 7: Error between test input u and reconstructed input u_r

4   Network models

In case of a circuit description of the chaotic transmission system differentiator blocks could be avoided. Network models have a mathematical description in terms of *differential algebraic equations (DAEs)*

$$\mathbf{F}\left(\dot{x}(t), \mathbf{x}(t), t\right) = 0.$$

Nevertheless, if the transmitter of a chaotic transmission systems had a relative degree $r > 0$, then the circuit equations of the receiver would have an index $\mu \geq r + 1$, and the numerical simulation would become difficult, see [7, 8, 33–37]. Obviously, the same problems would arise if the block-oriented simulator is just a front-end of a general network simulator (e.g. KOSIM [38]). Due to the use of conventional integration methods (e.g. implicit Euler) many simulators are not able to cope with higher index problems. One approach to circumvent these problems is to include derivatives of $\mathbf{F}$ into the computation. The *index* of a DAE is defined as the number of derivatives required to obtain an explicit ODE [33, 34, 39–42]. Hence, our method to compute Taylor coefficients in block simulation schemes is very similar to an index reduction procedure of network simulators.

5 Conclusions

A new method for the numerical simulation of systems with differentiator blocks has been introduced. This approach is based on automatic differentiation techniques. The feasibility has been demonstrated on a chaotic transmission system. Further research will be devoted to control-theoretic applications [43–46].

6 Acknowledgement

Support by the Deutsche Forschungsgemeinschaft is gratefully acknowledged. Udo Ließ and Alan Lynch are thanked for many interesting discussions which have contributed to the results presented.

7 References

[1] A. Isidori. *Nonlinear Control Systems: An Introduction.* Springer-Verlag, 3 edition, 1995.

[2] H. Nijmeijer and A. J. van der Schaft. *Nonlinear Dynamical Control systems.* Springer Verlag, 1990.

[3] M. Fliess, J. Lévine, P. Martin, and P. Rouchon. Flatness and defect of non-linear systems: Introductory theory and examples. *Int. J. Control,* 61:1327–1361, 1995.

[4] J. Rudolph, editor. *Fortbildungskurs Flachheitsbasierte Regelung,* Institut für Regelungs- und Steuerungstheorie, TU Dresden, 1997.

[5] R. Rothfuß, J. Rudolph, and M. Zeitz. Flachheit: Ein neuer Zugang zur Steuerung und Regelung nichtlinearer Systeme. *Automatisierungstechnik,* 45:517–525, 1997.

[6] J. v. Löwis, J. Rudolph, and K. J. Reinschke. Neues Regelungskonzept für Sensor-Aktor-Systeme: Lageregelung eines Langhubmagneten. Technical Report SFB 358-D5-1/98, TU Dresden, Sonderforschungsbereich 358, 1998.

[7] U. Feldmann, M. Hasler, and W. Schwarz. Communication by chaotic signals: the inverse system approach. *Int. J. of Circuit Theory and Applications*, 24(5):551–579, 1996.

[8] U. Feldmann, M. Hasler, and W. Schwarz. On the design of a synchronizing inverse of a chaotic system. In *Proc. Europ. Conf. Circ. Th. Design (ECCTD)*, Istanbul, Turkey, volume 1, pages 479–482, 1995.

[9] O. E. Rössler. An equation for continuous chaos. *Phys. Lett.*, 57A:397, 1976.

[10] O. E. Rössler. Continous chaos — four prototyp equations. In O. Gruel and O. E. Rössler, editors, *Bifurcation Theory and Applications in Scientific Disciplines*, volume 316 of *Ann. N.Y. Acad. Sci.*, pages 376–392, 1979.

[11] G. Jetschke. *Mathematik der Selbstorganisation*. VEB Deutscher Verlag der Wissenschaften, Berlin, 1989.

[12] H. G. Schuster. *Deterministic Chaos - An Introduction*. VCH Verlagsgesellschaft mbH, Weinheim, 1989.

[13] B. Speelpenning. *Compiling Fast Partial Derivatives of Functions Given by Algorithms*. PhD thesis, Department of Computer Science, University of Illinois at Urbana-Champaign, Urbana-Champaign, Ill., Jan. 1980.

[14] G. F. Corliss. Automatic differentiation bibliography. In A. Griewank and G. F. Corliss, editors, *Automatic Differentiation of Algorithms: Theory, Implementation, and Application*, pages 331–353. SIAM, Philadelphia, Penn., 1991.

[15] G. Corliss. Automatic differentiation bibliography. Technical Memorandum ANL/MCS–TM–167, Mathematics and Computer Science Division, Argonne National Laboratory, Argonne, Ill., July 1992.

[16] A. Griewank and G. F. Corliss, editors. *Issus in Parallel Automatic Differentiation*. 1991.

[17] M. Iri. History of automatic differentiation and rounding error estimation. In A. Griewank and G. F. Corliss, editors, *Automatic Differentiation of algorithms: Theory, implementation, and application*, pages 3–16, Breckenridge, Colorado, 1991. Proc. of the 1st SIAM Workshop on Automatic Differentiation.

[18] B. Christianson. Reverse accumulation and accurate rounding error estimates for Taylor series. *Optimization Methods and Software*, 1:81–94, 1992.

[19] A. Griewank. On automatic differentiation. In M. Iri and K. Tanabe, editors, *Mathematical Programming: Recent Developments and Applications*, pages 83–108. Kluwer Academic Publishers, 1989.

[20] A. Griewank. *Evaluating Derivatives — Principles and Techniques of Algorithmic Differentiation*, volume 19 of *Frontiers in Applied Mathematics*. SIAM, 2000.

[21] A. Griewank, D. Juedes, and J. Utke. A package for automatic differentiation of algorithms written in C/C++. *ACM Trans. Math. Software*, 22:131–167, 1996. http://www.math.tu-dresden.de/wir/project/adolc/index.html.

[22] C. Bendtsen and O. Stauning. TADIFF, a flexible C++ package for automatic differentiation. Technical Report IMM-REP-1997-07, TU of Denmark, Dept. of Mathematical Modelling, Lungby, 1997.

[23] M. Grundmann and M. Masmoudi. MIPAD: an AD package. AD'2000, Nice (France), 2000.

[24] D. Werner. *Funktionalanalysis*. Springer-Verlag, 1997.

[25] T. F. Coleman and A. Verma. Structured automatic differentiation. http://www.cs.cornell.edu/Info/People/verma/AD/research.html.

[26] T. F. Coleman and A. Verma. ADMIT-1: Automatic differentiation and MATLAB interface toolbox, user guide, release alpha 1. Technical Report CTC97TR271, Cornell Theoy Center, 1997.

[27] T. F. Coleman and A. Verma. ADMIT-1: Automatic differentiation and MATLAB interface toolbox. Available as Cornell Computer Science technical report CS TR 98-1663 and as Cornell CCOP Technical report TR 98-1, 1998.

[28] A. Verma. ADMAT: Automatic differentiation for MATLAB using object oriented methods. Submitted to SIAM workshop on object oriented methods, 1999.

[29] A. Griewank. ODE solving via automatic differentiation and rational prediction. In D. F. Griffiths and G. A. Watson, editors, *Numerical Analysis 1995*, volume 344 of *Pitman Research Notes in Mathematics Series*. Addison-Wesley, 1995.

[30] G. F. Corliss and Y. F. Chang. Solving ordinary differential equations using Taylor series. *ACM Trans. Math. Software*, 8:114–144, 1982.

[31] Y. F. Chang. Automatic solution of differential equations. In D. L. Colton and R. P. Gilbert, editors, *Constructive and Computational Methods for Differential and Integral Equations*, volume 430 of *Lecture Notes in Mathematics*, pages 61–94. Springer Verlag, New York, 1974.

[32] Y. F. Chang. The ATOMCC toolbox. *BYTE*, 11(4):215–224, 1986.

[33] K. E. Brenan, S. L. Campbell, and L. R. Petzold. *Numerical Solution of Initial-Value Problems in Differential-Algebraic Equations*. North-Holland, 1989.

[34] E. Griepentrog. The index of differential-algebraic equations and its significance for the circuit simulation. In *Mathematical Modelling and Simulation of Electrical Circuits and Semiconductor Devices, Conference held at the Mathematisches Forschungsinstitut, Oberwolfach*, pages 11–25. Birkhäuser Verlag, 1990.

[35] E. Griepentrog and R. März. *Differential-Algebraic Equations and Their Numerical Treatment*, volume 88 of *Teubner-Texte zur Mathematik*. Teubner Verlagsgesellschaft, Leipzig, 1986.

[36] E. Hairer, C. Lubich, and M. Roche. *The Numerical Solution of Differential-Algebraic Systems by Runge-Kutta Methods*, volume 1409 of *Lecture Notes in Mathematics*. Springer-Verlag, 1989.

[37] B. Straube, K. Reinschke, W. Vermeiren, K. Röbenack, B. Müller, and C. Clauß. DAE-index increase in analogue fault simulation. In *Proc. Workshop on System Design Automation SDA'2000, Rathen (Germany), March 13-14*, pages 99–106, 2000.

[38] P. Schwarz et al. KOSIM — ein Mixed-Mode, Multi-Level-Simulator. *Informatik-Fachberichte*, 225:207–220, 1990.

[39] C. W. Gear and L. R. Petzold. ODE methods for the solution of differential/algebraic systems. *SIAM J. Numer. Anal.*, 21(4):717–728, August 1984.

[40] C. W. Gear. Differential-algebraic equation index transformations. *SIAM J. Sci. Stat. Comput.*, 9(1):39–47, 1988.

[41] P. J. Rabier and W. C. Rheinboldt. A general existence and uniqueness theory for implicit differential-algebraic equations. *Differential and Integral Equations*, 4(3):563–582, May 1991.

[42] S. Reich. On a geometrical interpretation of differential-algebraic equations. *Circuits Systems Signal Processing*, 9(4):367–382, 1990.

[43] K. Röbenack. Nutzung des Automatischen Differenzierens in der nichtlinearen Regelungstheorie. In *Gemeinsamer Workshop des GAMM-Fachausschusses "Dynamik und Regelungstheorie" und des VDI/VDE-GMA-Ausschusses 1.40 "Theoretische Verfahren der Regelungstechnik", Kassel, 01.-02. März*, 1999.

[44] K. Röbenack and K. J. Reinschke. Trajektorienplanung und Systeminversion mit Hilfe des Automatischen Differenzierens. In *Workshop des GMA-Ausschusses 1.4 "Neuere theoretische Verfahren der Regelungstechnik", Thun, Sept. 26-29*, pages 232–242, 1999.

[45] K. Röbenack and K. J. Reinschke. Reglerentwurf mit Hilfe des Automatischen Differenzierens. *Automatisierungstechnik*, 48(2):60–66, 2000.

[46] K. Röbenack and K. J. Reinschke. Nonlinear observer design using automatic differentiation. AD'2000, Nice, 2000.

DAE-Index Increase in Analogue Fault Simulation[1]

Bernd Straube[*], Kurt Reinschke[**], Wolfgang Vermeiren[*], Klaus Röbenack[**], Bert Müller[*], Christoph Clauß[*]

[*] Fraunhofer-Institut für Integrierte Schaltungen, Erlangen,
Design Automation Department EAS Dresden

[**] Technische Universität Dresden, Fakultät Elektrotechnik,
Institut für Regelungs- und Steuerungstheorie

Abstract

In this paper, we investigate the importance of DAE-index in analogue fault simulation. The authors show that high indices which result from fault injection can lead to problems in numerical simulation. Methods for index computation which can be appended to standard fault detection schemes allow the prediction of potential numerical instabilites. Some suggestions to tackle such index problems will be discussed. Examples of the use of our index approach are given.

1   Introduction

Let us consider an arbitrary analogue circuit. The proper function of this circuit as well as its behaviour in case of faults (e.g. *shorts* and *opens*) is usually analysed through numerical simulations based on network models. Circuit simulators such as SPICE model these networks by *differential-algebraic equations (DAEs)*. In contrast to explicit *ordinary differential equation* (ODEs), DAEs may additionally contain algebraic equations, e.g. the equations resulting from Kirchhoff's voltage and current laws.

DAEs are a more general system description than ODEs, i.e., there are systems that do not have an ODE representation but can be written as a DAE. Furthermore, many approaches to modelling physical systems result in a DAE, e.g. the modified nodal analysis in case of electrical networks or the equations of motion for mechanical systems. On the other hand, DAEs can exhibit mathematical phenomena which are not possible for ODEs. For example, the solution of DAEs may depend on derivatives of the input signal. To classify the problems associated with DAEs, the concept of the *DAE index* has been introduced. Even though there are various index definitions, most of these definitions coincide with each other in the case of linear time-invariant systems.

The stability and other analytical properties of numerical DAE solvers depend essentially on the index [4, 2, 8]. DAEs with an index zero or one can be solved easily with standard ODE methods. However, higher index problems (i.e., DAEs with an index greater than one) constitute serious problems to numerical solvers. Since the solution of a DAE can depend on time derivatives of the input signal, small perturbations of this input signal may cause arbitrary large errors of the solution [8]. Moreover, the index is not given by the electrical network itself but depends on the kind of equations [5, 6, 7, 12].

---

[1] *This work has been supported by the Deutsche Forschungsgemeinschaft, Sonderforschungsbereich 358*

R. Merker and W. Schwarz (eds.), System Design Automation, 221-232.

The aim of an analogue fault simulation is to derive appropriate test signals for the real circuit. Details related to analogue fault simulators are discussed elsewhere [22, 18, 19, 1, 21, 3, 17]. In this paper we will only discuss the effect of an increase in the index caused by fault injections.

Let us consider an analogue network. The design process is supposed to be finished, i.e., this network meets all specified requirements. Furthermore, due to the applied modelling techniques the network has a small index (i.e., zero or one) such that no problems will arise during the numerical simulations.

A fault simulator creates a faulty network by fault injection. The electrical behaviour of a fault is usually modelled by network elements and their interconnections. The most commonly injected faults are *shorts* and *opens*. Injecting such faults can dramatically alter the circuit's behaviour. For the given input signals the deviations of the output signals of the faulty network from the fault-free one may be small or large. However, the faulty network may exhibit unbounded deviations. This is regarded as an instability of the defective circuit. In this case the analogue fault simulator will mark this fault as to be detected.

The purpose of this paper is to point out that the index of the underlying system of differential-algebraic equations may increase when a fault is injected. Then, the following situations could arise. The network simulator

(1) is able to calculate the solution properly,

(2) tries repeatedly to find a new time step size without success,

(3) aborts the simulation with an error message,

(4) calculates unbounded output signals or other completely false outputs.

The cases (2) and (3) are managed by the control module of the analogue fault simulator without any problems [21]. In this case, a fault which causes this instability is marked as *non-simulated* and, therefore, it could *not be detected*. However, in case (4) the evaluation module of the fault simulator is unable to distinguish between an actual „electrical" instability and a merely „numeric" one, i.e., between a proper deviation caused by the fault or a numerical instability caused by a large index. Hence, the fault is marked as detected as soon as the output signal of the faulty network is outside the specified pass-band. Since the actual behaviour of this fault remains unknown this test signal may lead to a potential *test escape*.

2   The index of linear network equations

In this section we outline one concept of a system's index. Linear electrical networks are described by linear DAEs

$$A\frac{d}{dt}x(t) + Bx(t) + f(t) = 0 \tag{1}$$

$(x(t), f(t): T \to R^m, A, B \in R^{m \times m}, T \subseteq R^1 \text{ time interval, } m > 0)$

The matrices A and B as well as their dimensions depend on the choice of the variables. For example, the *branch-voltage branch-current equations* have the form depicted in Fig. 1 where $v_1, ..., v_b$ and $i_1, ..., i_b$ denote the voltages and the currents of the b network branches, respectively. The matrix K results from the incidence matrices describing the network topology or, in other words, the Kirchhoff's voltage and current laws. The *v-i-relations* of the resistors are represented by the matrix M. The matrices F_{LC} and H_{LC} represent the *v-i-relations* of the capacitors and inductors.

$$\begin{bmatrix} 0 \\ \hline 0 \\ \hline F_{LC} \end{bmatrix} \frac{d}{dt} \begin{bmatrix} v_1 \\ \vdots \\ v_b \\ i_1 \\ \vdots \\ i_b \end{bmatrix} + \begin{bmatrix} K \\ \hline M \\ \hline H_{LC} \end{bmatrix} \begin{bmatrix} v_1 \\ \vdots \\ v_b \\ i_1 \\ \vdots \\ i_b \end{bmatrix} + f(t) = 0$$

$$A\frac{d}{dt}x(t) + Bx(t) + f(t) = 0$$

Fig. 1: System of branch-voltage branch-current equations

Next, the computation of the tractability index will be discussed. This computation can not only be used to check whether the DAE of a given networks has a higher index but also to find out how parameters of the network elements have to be chosen to avoid higher indices.
According to the tractability index definition [11] the following matrix chains have to be calculated,

$$A_0 = A \tag{2}$$

$$B_0 = B \tag{3}$$

$$A_{i+1} = A_i + B_iQ_i \tag{4}$$

$$B_{i+1} = B_iP_i \tag{5}$$

where Q_i is a projector on the kernel $ker\,A_i$ and $P_i = I - Q_i$ for $i = 0,1,...,m$. The tractability index is the smallest number τ for which A_τ is regular. Consequently, the index is higher than one if and only if A_1 is singular. Therefore, the condition

$$det(A + BQ_0) = 0 \tag{6}$$

has to be checked. To get this condition symbolic network analysis is used.
Analog Insydes [9] is a program for the symbolical analysis of analogue networks. It is based on the computer algebra system Mathematica [23] and utilizes its symbolic calculation capabilities. For handling network analysis a set of Mathematica functions is defined. We use Analog Insydes to derive (1) from a network description. The further calculations use Mathematica functions only. The higher index condition (6) is derived by the following steps:

• Describe the network in an element oriented way. (There is an input format which uses the Mathematica list capabilities. A netlist converter for reading SPICE nudists is available).

• Generate the network equations. (It is possible using the'CircuitEquations' analysis function with the 'SparseTableau' option.)

• Extract the A and B matrices. (This is possible using Mathematica functions.)

• Construct the projector Q_0 on $ker\,A$ with sparse matrix techniques.

• Calculate the determinant of $A_1 = A + BQ_0$. (Symbolic determinant calculation and factorization are functions of the Mathematica system.)

The symbolic analysis steps are combined into a Mathematica function.

As mentioned above there are also other definitions of the index. Furthermore, there are efficient methods to compute the structural index based on the network topology or on the sparsity pattern of the matrices A and B alone. More details and further investigations of the index are found elsewhere [10, 6, 14, 16, 13, 15].

3 Three simple examples for an increasing index caused by fault injection

3.1 Example 1: RC-network with an operational amplifier

The first example to be considered is the RC-network with an operational amplifier in Fig. 2. If the operational amplifier is modelled ideally by a nullator-norator-pair the fault free network can be simulated without any problems.

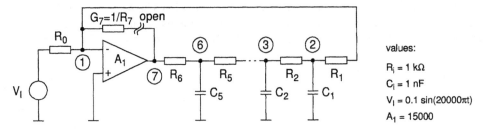

Fig. 2: RC-network using a voltage-controlled voltage source as op-amp

The behaviour (node voltage 7) of the fault-free network is shown in Fig. 3.

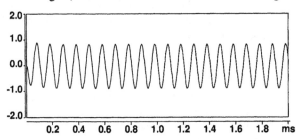

Fig. 3: Behaviour of the network of Fig. 2 (node voltage 7)

An *open* fault is injected in series with resistor R_7. In the case of this open fault no simulation was possible. Otherwise, if the operational amplifier is modelled by using a voltage controlled voltage source (amplification $V = 15000$) the fault-free network can also be simulated. The output voltage (node 7) does not differ visually from that in Fig. 3. If the open fault is injected at R_7, the simulator has calculated the output shown in Fig. 4.

However, the DAE of this network has the *index = 6*. This is the reason for the numerically computed unbounded output shown in Fig. 4. An analogue fault simulator would report this fault as to be detected.

Practically, it is impossible to get knowledge about the actual behaviour of the faulty RC-network by means of numerical simulation.

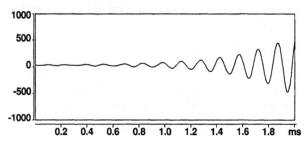

Fig. 4: Behaviour of the network of Fig. 2 with the injected *open* at the resistor R_7, voltage at node 7 (*Note the different voltage ranges of the signals*)

Since the considered network is linear the analytical solutions can be found by means of symbolic analysis [9]. The output signal of the faulty network is

$$V_7 = \frac{a_0+a_1s+a_2s^2+a_3s^3+a_4s^4+a_5s^5}{b_0+s^2}$$

with the coefficients

$a_0 = -3.7699 \ 10^3$ $a_1 = -5.6712$

$a_2 = -1.5100 \ 10^{-5}$ $a_3 = -1.3832 \ 10^{-11}$

$a_4 = -5.0278 \ 10^{-18}$ $a_5 = -6.2830 \ 10^{-25}$

$b_0 = 4.0\pi^2 \ 10^{-8}$

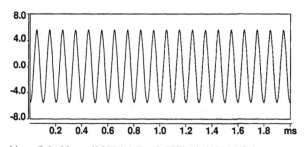

$V_7 = -5.6166 \cos(62831.9 \ t) + 0.8875 \sin(62831.9 \ t)$

Fig. 5: Analytical solution for the voltage at node 7 of the network of Fig. 2 with the injected *open* fault at resistor R_7

Since the degree of the nominator polynomial is higher than the degree of the denominator polynomial, the system must have an *index > 1*. Furthermore it can be seen that any input signal is differentiated. The analytical solution which differs from the simulation results drastically (Fig. 4) is depicted in Fig. 5.

The application of the higher index condition (6) allows

• detection of a higher index immediately after fault injection but without the fault simulation run,

• a warning if in the case of a higher index a fault is reported as to be detected.

On the other hand, the conditions for an index increase can be calculated for different description levels of the operational amplifier. In case of an ideal operational amplifier we get the following condition for a higher index

$$C_1 C_2 C_3 C_4 C_5 G_7 R_0 R_1 R_2 R_3 R_4 R_5 R_6 = 0 \qquad (7)$$

That means the index increases if $G_7 = 0$ which is the *open* fault, or if $R_i = 0$ ($i = 0, ..., 6$) which leads to loops of capacitors. If the operational amplifier is represented by a voltage controlled voltage source (amplification V) the higher index condition changes into

$$C_1 \cdot ... \cdot C_5 R_2 \cdot ... \cdot R_6 (R_0 + R_1 + G_7 R_0 R_1 - G_7 R_0 R_1 V) = 0 \qquad (8)$$

The index increases if:

$$G_7 = (R_0 + R_1)/(R_0 R_1 (V-1)) = 0 \qquad (9)$$

In the case of the open fault the network can be simulated because the index is not high. But the result is doubtful because G_7 is in the critical region according to (9) if G_7 comes closer to zero. The symbolic higher index condition gives a useful insight.

3.2   Example 2: High-pass filter

The second example is a high-pass filter with *index = 1* (Fig. 6).

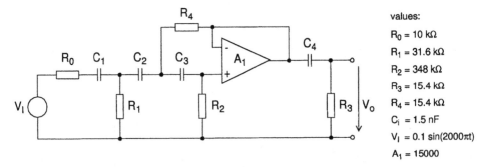

values:
$R_0 = 10 \text{ k}\Omega$
$R_1 = 31.6 \text{ k}\Omega$
$R_2 = 348 \text{ k}\Omega$
$R_3 = 15.4 \text{ k}\Omega$
$R_4 = 15.4 \text{ k}\Omega$
$C_i = 1.5 \text{ nF}$
$V_I = 0.1 \sin(2000\pi t)$
$A_1 = 15000$

Fig. 6: High-pass filter using a voltage-controlled voltage source as op-amp.
Short fault at R_4, $R_{short} = 10\Omega$

The output voltage V_o of the fault-free network is shown in Fig. 7.

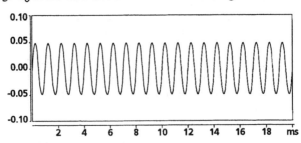

Fig. 7: Output voltage V_o of the high-pass filter of Fig. 6

A *short* fault is injected at resistor R_4. This causes a higher index. The simulator has calculated the output shown in Fig. 8.

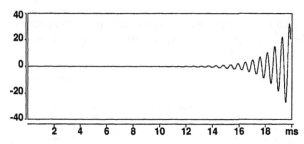

Fig. 8: Output voltage V_o of the high-pass filter of Fig. 6 with the injected
short at the resistor R_4

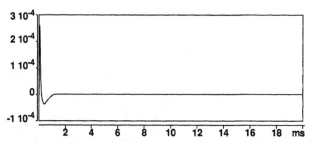

$V_o = -0.0004 \; 10^{-1.341t} +0.0006 \; 10^{-161668t} -7.6573 \; 10^{-7}\cos(6283.19 \; t) + 8.0332 \; 10^{-7}\sin(6283.19 \; t)$

Fig. 9: Analytical solution for the voltage V_o of the high-pass filter of Fig. 6
with the injected *short* at resistor R_4

Again, the analytical solution was calculated by means of symbolic analysis. The transfer function for the faulty network is

$$\frac{V_o}{V_I} = \frac{a_4 s^4}{(c+ds)(b_0+b_1s+b_2s^2+b_3s^3)}$$

with the coefficients

$$a_4 = 1.2860 \qquad c = 1.0 \; 10^7 \qquad d = 2.31 \; 10^2$$
$$b_0 = 1.5001 \; 10^{13} \qquad b_1 = 1.6481 \; 10^9$$
$$b_2 = 1.0879 \; 10^4 \qquad b_3 = 7.8605 \; 10^{-3}$$

In contrast to the simulated behaviour (Fig. 8) the analytical solution is completely different (Fig. 9).
In this example, the network equations derived from an ideal operational amplifier model have a higher index if the condition

$$C_1C_2C_3C_4(R_0R_1+R_0R_2+R_1R_2)R_3R_4 = 0 \tag{10}$$

is fulfilled. Hence, the index becomes two if we insert a *short* fault by $R_4 = 0$.

3.3   Example 3: Chaotic Network

Fig. 10 shows the signal flow of a chaotic transmission system. Transmitter and Receiver have been depicted in Fig. 11 and Fig. 12, respectively. Both chaotic systems are based on Chua's circuit [27] as sketched in Fig. 13.

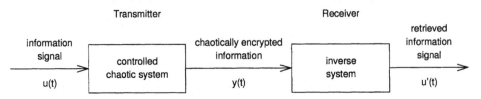

Fig. 10: Chaotically encrypted signal transmission

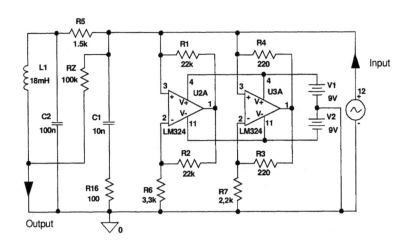

Fig. 11: Chaotic transmitter circuit

The transmitter can be modelled by a state-space system with the relative degree zero (see [28]). The receiver is designed according to the inverse system approach [24, 25, 26]. The circuit equations of the receiver have the *index = 1*. The numerical simulation is uncritical: Fig. 14 shows an periodical input signal, the chaotically encrypted signal and the reconstructed signal on the output of the receiver.

Now, we will analyse the behaviour of this transmission system in case of an *open* fault which has been injected on the resistor R_Z. In this case, the index of the network equations of the receiver increases. The *branch voltage branch current equations* as well as the modified nodal analysis equations have *index = 3*. (For the index calculation, it is useful to consider parts of the circuit as a nonlinear negative resistor. The subcircuit is shown in the gray shaded box in Fig. 12.) In simulation, the open fault was modelled by $R_{open} = 10M\Omega$. The simulator tried to find a suitable step size numerous times without success. Eventually, the simulator had to be interrupted.

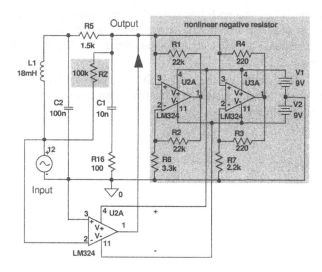

Fig. 12: Chaotic receiver circuit

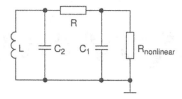

Fig. 13: Chua's circuit

Further examples of index changes due to fault injection can be found in [20].

4 Possible measures for tackling the index problem

- Fault injection may undo the designer's index-related modelling measures. Based on designer's knowledge the index-related elements ought to be labelled in the fault-free network. An increase of the index could be avoided if these labelled elements were not set faulty or the corresponding faults were deleted from the fault list.

- A numerical supervision of the index [13] is very time-consuming. However, the index should be estimated when an output signal leaves the pass- band. A simple method for the symbolic calculation of the higher index condition for linear networks has been presented. For large scale systems we propose graph-theoretic methods to compute the structural index [14].

- As the kernel of the analogue fault simulator we propose the use of an electric simulator which interrupts the simulation of an index-increasing fault in case of numerical instabilities.

In this way it could ensure that such faults are classified as non-simulated. Therefore, it is necessary to define analogue benchmark networks with some index-increasing faults. By means of these benchmarks a given electrical simulator can be checked whether it is suitable as a fault simulator.

• The dependence of the index on the used fault simulation models (e.g. resistor model, replacement model, zero-source model), the fault sites, the simulation modes, and the input signals have to be investigated.

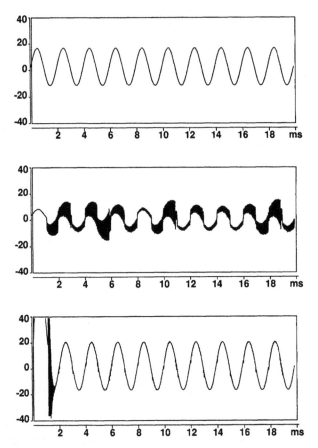

Fig. 14: The fault-free behaviours of the transmission line (Fig. 10): input
of the transmitter (above), transmitted signal (middle), output of the
receiver (below)

5   Conclusion

In practice, carrying out analogue fault simulations takes a lot of time and it is not possible to supervise the complete fault simulation process. Therefore, it is important that dependable simulation results are provided by the fault simulation tool.

In this paper we have described networks that show no numerical problems during their fault-free simulations but may have increased their indices due to the injection of faults. The higher index causes numerical problems during simulation. An analogue fault simulator is able to react if the electric simulator stops its calculation somehow. However, when the faulty behaviour leaves the pass-band, the fault simulator cannot distinguish between the actual faulty behaviour and numerical instabilities of the system of differential-algebraic equations. This results in a false detection classification of the fault (test escape). From the fault simulation point of view the index-caused problem remains unsolved although some practical remedies have been described.

6   References

[1] I. M. Bell, S. J. Spinks: Analogue Fault Simulation for the Structural Approach to Analogue and Mixed-signal IC Testing. International Mixed Signal Testing Workshop, June 20-22, 1995, Grenoble/Villard de Lans, France. Collection of Papers, pp. 10-14.

[2] K. E. Brenan, S. L. Campbell, L. R. Petzold: Numerical Solution of Initial-Value Problems in Differential-Algebraic Equations. North-Holland, 1989.

[3] P. Caunegre, C. Abraham: Fault Simulation for Mixed-Signal Systems. Journal of Electronic Testing: Theory and Application (JETTA), vol. 8 (1996) No.2, April 1996, pp.143-152.

[4] E. Griepentrog, R. März: Differential-Algebraic Equations and Their Numerical Treatment. Band 88 aus Teuber-Texte zur Mathematik, Teubner Verlagsgesellschaft, Leipzig, 1986.

[5] M. Günther, U. Feldmann: The DAE-Index in Electric Circuit Simulation. Proc. IMACS Symp. Mathematical Modelling. (Ed. I. Troch, F. Breitenecker) vol. 4, pp. 695-702, 1994

[6] M. Günther, U. Feldmann: CAD based electric circuit modeling. Part I: Mathematical structure and index of network equations. Surv. Math. Ind. (1999) 8, pp. 97-129

[7] M. Günther, U. Feldmann: CAD based electric circuit modeling. Part II: Impact of network structure and parameters. Surv. Math. Ind. (1999) 8, pp. 131-157

[8] E. Hairer, Ch. Lubich, M. Roche: The Numerical Solution of Differential-Algebraic Systems by Runge-Kutta Methods, Band 1409 aus Lecture Notes in Mathematics, Springer-Verlag, 1989.

[9] Hennig, E. and T. Halfmann. 1998. Analog Insydes Tutorial. ITWM, Kaiserslautern, Germany

[10] W. Kampowsky, P. Rentrop, W. Schmidt: Classification and Numerical Simulation of Electric Circuits. Surveys on Mathematics for Industry (1992) 2, pp. 23-65.

[11] März, R., Numerical methods for differential-algebraic equations. Acta Numerica, 1991, pp. 141-198.

[12] K. Matz, C. Clauß: Zur Simulationspraxis bei DAE's mit höherem Index. 4. GMM/ITG Diskussionssitzung - Entwicklung von Analogschaltungen mit CAE-Methoden, Berlin, Oktober 1997, pp. 101 - 108

[13] K. Matz, C. Clauß: Simulation Support by Index Computation. 15th IMACS World Congress, Berlin, Aug. 1997

[14] K. Reinschke, K. Röbenack: Graphentheoretische Bestimmung des strukturellen Index von Algebro-Differential-Gleichungssystemen für die Netzwerkanalyse. In W. Mathis, P. Noll (Hrsg.): Neue Anwendungen theoretischer Konzepte in der Elektrotechnik - Tagungsberichte der 2. ITG-Diskussionssitzung vom 21.-22. April 1995 in Berlin, pp. 207-214, VDE-Verlag, 1996.

[15] K. Röbenack, K. Reinschke: Graph-theoretically determined Jordan block size structure of regular matrix pencils, Linear Algebra Appl. 263 (1997), pp. 333-348.

[16] G. Reißig, U. Feldmann: Computing the generic index of the circuit equations of linear active networks. In Proc. Int. Symp. on Circuits and Systems (ISCAS), Altanta, 12.-15. Mai 1996, Band III, pp. 190-193.

[17] B. Straube, W. Vermeiren, B. Müller, Ch. Hoffmann, S. Sattler: Analoge Fehlersimulation mit aFSIM. Analog'99, 5. ITG/GMM-Diskussionssitzung 'Entwicklung von Analogschaltungen mit CAE-Methoden', Munich, Februar 1999

[18] M. Sachdev: A Defect Oriented Testability Methodology for Analog Circuits. Journal of Electronic Testing: Theory and Applications (JETTA) vol. 6 (1995) no. 3 (June), pp. 265-276.

[19] Chr. Sebeke, J. P. Teixeira, M. J. Ohletz: Automatic Fault Extraction and Simulation of Layout Realistic Faults for Integrated Analogue Circuits. European Design and Test Conference 1995, Paris, France, March 6 - 9, 1995, pp. 464-468.

[20] B. Straube, K. Reinschke, W. Vermeiren, K. Röbenack: Indexveränderung durch Fehlerinjektion. Technical Report SFB 358-C1D5-1/97, TU Dresden

[21] B. Straube, W. Vermeiren, A. Holubek, M. J. Ohletz, M. Dhifi: Analogue Fault Simulation Tools aFSIM and AnaFAULT. 8. ITG/GI/GME-Workshop „Testmethoden und Zuverlässigkeit von Schaltungen und Systemen", Otzenhausen, 3.- 5. März 1996, pp. 78 - 80.

[22] W. Vermeiren, B. Straube, G. Elst: A Suggestion for Accelerating the Analogue Fault Simulation. In: Proc. EDAC-ETC-EUROASIC 1994, Paris, France, Feb. 28 - March 3, p. 662, IEEE Comp. Society Press, Los Alamitos, CA, 1994

[23] Wolfram, S., Mathematica - Ein System für Mathematik auf dem Computer. Addison-Wesley Publishing Company, 1992.

[24] U. Feldmann, M. Hasler: Inverse System Realisation with Operational Amplifier - Stability vs. nonideal Op.-Amp. characteristics. NDES'95, 1995, pp. 139-142.

[25] U. Feldmann, M. Hasler, W. Schwarz: On the design of a synchronizing inverse of a chaotic system. ECCTD'95, Istanbul, 1995, pp. 479-482.

[26] U. Feldmann, M. Hasler, W. Schwarz: Communication by Chaotic Systems: The Inverse Systems Approach. Int. J. Circuit Theory and Applications 24 (1996) 5, pp. 531-579.

[27] L. O. Chua: The genesis of Chua's circuit. Archiv für Elektrotechnik und Übertragungstechnik 46 (1992) 4, pp. 250-257.

[28] A. Isidori: Nonlinear Control Systems. 3. Edition, Springer-Verlag, 1995.

Parameter Optimization of Complex Simulation Models

Dipl.-Inform. Elisabeth Syrjakow, Institute for Computer Design and Fault Tolerance (Prof. D. Schmid), University of Karlsruhe, 76128 Karlsruhe, Germany

Dr.-Ing. Michael Syrjakow, Institute for Computer Design and Fault Tolerance (Prof. D. Schmid), University of Karlsruhe, 76128 Karlsruhe, Germany

Abstract

Today, a great shortcoming of most of the available simulation tools is, that optimization is not sufficiently supported by appropriate optimization techniques. This situation is mainly caused by the lack of qualified strategies for optimization of simulation models. Traditional indirect optimization techniques based on exploitation of analytical information like gradients etc. cannot be applied, because only goal function values are available which have to be calculated by an often very expensive simulation process. In the first part of this paper, some powerful direct optimization algorithms are presented which work iteratively, only requiring goal function values. These strategies are combinations of direct global and local optimization methods (Genetic Algorithms, Simulated Annealing, Hill-Climbing, etc.) trying to merge the advantages of global and local search. Our developed strategies have been implemented and integrated into REMO (REsearch Model Optimization package) representing a software tool for experimentation with simulation models. At the moment REMO is extended and completely reimplemented in Java to make it available on the World Wide Web. The ongoing development process of REMO is described in the second part of this paper.

1   Introduction

The global goal of system design is to build optimal systems regarding costs, performance, and reliability. Consequently, the development of appropriate optimization methods is a rewarding and profitable research field in mathematics and also in computer science.

Unfortunately, because of the increasing system complexity, it becomes more and more impossible to formulate the goal function, which has to be optimized by mathematical means. The reason for this insufficiency is the lack of appropriate analytical modelling techniques allowing to describe complex system structures and their dynamic behaviour in that way, that the desired performance and reliability measures can be derived mathematically. For that reason, often no traditional analytical optimization methods can be applied and the one and only way is to optimize by making experiments on the real system itself or on a simulation model of it.

In most cases, experimentation with the real system itself is rather difficult or even impossible. Reasons for that may be:

- the real system does not exist, because it is still in planning or development,
- the real system is difficult to access,
- experiments on the real system are inadmissible impairments,
- experiments on the real system are very dangerous,

R. Merker and W. Schwarz (eds.), System Design Automation, 233-246.

- experiments on the real system are expensive, uneconomical, and taking long periods of time,
- state transitions of the real system occur too slowly or too rapidly.

For that reason, in many cases the only way to succeed in system optimization is to employ simulation models. On the one hand, simulation models are very easy to modify and to manipulate. On the other hand, they are abstract representations of the real system, comprising only relevant properties for the optimization process. Hence, simulation models play an entirely significant role in system optimization.

Having built a system model, the task of model optimization is to find out model input parameter values, causing optimal system model behaviour or at least an improvement of it. Subsequently, analysis and interpretation of the optimized system model may deliver essential hints for an optimal design or tuning of the real system.

In order to automate the process of model optimization the software tool REMO (REsearch Model Optimization package) was developed [11]. To resolve the difficult task of model optimization REMO applies so-called direct optimization strategies, which work iteratively only requiring goal function values. Today the most common direct methods for global optimization are Genetic Algorithms [3,6], Evolution Strategies [8], and Simulated Annealing [1]. All these methods are based on sophisticated probabilistic search operators, which imitate principles of nature. Although these operators have been proven to be well suited for global search, the required computational effort (number of goal function evaluations) as well as the quality of the optimization result still remains a big problem. In order to substantially improve the performance of direct optimization we have combined probabilistic global with deterministic local optimization methods. The excellent heuristic properties of this hybrid method make it possible to use it for multiple-stage optimization. Here the goal is not to localize only one but systematically the most important extreme points of a given optimization problem [10,12]. At the moment REMO is completely redesigned in order to make it a user-friendlier tool aimed at professional usage.

In the following section we will introduce into parameter optimization of simulation models. Section 3 contains a brief description of our developed optimization strategies, which are provided by the model optimization tool REMO. In Section 4 the current state as well as the ongoing development process of REMO is described. Finally, in Section 5, we summarize and draw some conclusions.

2 Parameter Optimization of Simulation Models

Each instance of an optimization problem can be formalized as a pair (S,F). The solution space S denotes the set of all possible problem solutions, which are valued by a goal function F being a mapping

$$F : S \rightarrow R. \tag{1}$$

The problem is to find a solution i* which satisfies

$$\forall i \in S : F(i) \circ F(i^*) = F^*, \quad \circ \in \{\leq, \geq\}. \tag{2}$$

A solution i^* is called a globally-optimal solution. The function value $F(i^*) = F^*$ is referred to as global optimum of F. Beside globally-optimal solutions there may exist locally-optimal solutions $i^\wedge$, having the property that all neighbouring solutions have the equal or a worse goal function value. Goal functions with several globally- and/or locally-optimal solutions are called multimodal functions. An optimization problem is either a minimization or a maximization problem. Minimization problems can be easily transformed into maximization problems and vice versa, because

$$\min(F(i)) = -\max(-F(i)). \tag{3}$$

Based on the general formulation of optimization problems above, we now describe the solution space S and the goal function F of our model optimization problem. As depicted in Fig. 1, a simulation model defines a relationship between model input parameters and model outputs, which are obtained by model evaluation through simulation using a simulation tool.

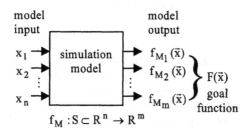

Fig. 1 Definition of a model-based goal function

For that reason, a simulation model can be viewed as a black-box function

$$f_M : S \subset R^n \rightarrow R^m \tag{4}$$

mapping a vector $\vec{x} = (x_1, x_2, ..., x_n)$, $x_i \in R$, $i \in \{1,..,n\}$ of model input parameters onto several model outputs $f_{M_j}(\vec{x}), j \in \{1,...,m\}$. From now on, a function f_M derived from a simulation model is referred to as model function. The model inputs can be roughly classified into system and workload parameters. The model outputs represent information about the system behaviour (performance, reliability, consumption of resources, etc.).

Obviously, optimization of a simulation model can be viewed as a parameter optimization problem. The problem solutions are parameter vectors $\vec{x}$. The goal is to find a vector $\vec{x}^*$ causing optimal model behaviour, which is quantified by a function F. F, representing the optimized goal function is composed of the model outputs $f_{M_j}(\vec{x})$, $j \in \{1,...,m\}$. The formulation of F may be very difficult, especially in case of contradictory system design goals. Frequently, F is defined as a weighted sum of model outputs

$$F(\vec{x}) = \sum_{k=1}^{m} \omega_k \cdot f_{M_k}(\vec{x}), \qquad \omega_k \in R. \tag{5}$$

Because F is based on a simulation model, it shows the following properties:

- expensive to evaluate
- stochastic inaccuracies in case of stochastic simulation models
- no additional analytical information (gradients, etc.) of any kind

All these characteristics and the fact, that global parameter optimization has proved to be NP-hard makes it extremely difficult to find high quality solution methods. Generally, intricate optimization problems from practice require solution methods based on heuristics. These methods have to fulfil the following requirements:

- exclusive utilization of goal function values
- suitability for global search
- fast and sure convergence
- high accuracy
- robustness against stochastic inaccuracies
- ability to cope with high numbers of parameters
- low implementation effort
- easy handling

Because only goal function values are available to guide the search process no indirect optimization strategies are applicable which exploit additional analytical information. This is a very hard constraint because the usually very expensive simulation process only allows to evaluate a very limited number of goal function values. Nevertheless, in this paper we will present some powerful direct optimization methods, which are able to cope with this demanding task.

To automate the process of model optimization a specific interface, shown in Fig. 2 is required. Its task is to couple the optimization with the simulation process, allowing data exchange as well as process synchronization. A vector of parameter values $\bar{x}$ generated by the direct optimization strategy is passed onto the simulation process. Then, a simulation tool evaluates the parameterised simulation model and sends back the calculated goal function value $F(\bar{x})$ to the optimization process. Outgoing from this value, the optimization strategy generates a new parameter vector, which is in turn transferred to the simulation process. That way, the goal function values are improved step by step until a termination condition is fulfilled.

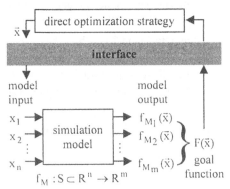

Fig. 2 The iterative process of model optimization

Note, that simulation models are abstract representations of real systems. Hence, before optimizing a simulation model, we have to assure that the model appropriately covers all relevant system properties. The results of a usually very expensive optimization process are of no use, if the model proves to be not validated sufficiently.

3   Optimization Strategies

To be able to cope with the non-trivial task of model optimization described above, we have developed a new kind of direct optimization algorithm called combined 2-phase strategy. The basic idea of this hybrid method is to split the optimization process into two phases: pre-optimization with a probabilistic global optimization method and fine-optimization performed by a deterministic local Hill-Climber. The task of pre-optimization is to explore the search space in order to find promising regions where global optimum points might be located. Outgoing from the promising region detected by pre-optimization the task of fine-optimization is to efficiently and exactly localize the optimum point in it. So, pre-optimization is pre-dominantly responsible for optimization success, whereas fine-optimization has to ensure the quality of the optimization result.

Fig. 3 shows the basic structure of a combined 2-phase optimization strategy os_{2P}. For realization of pre- and fine-optimization os_{2P} consists of a direct global and a direct local optimization method. The two strategies are coupled by means of an interface, comprising a method to derive control parameter values from optimization trajectories (dcp) as well as a method to select start points (ssp).

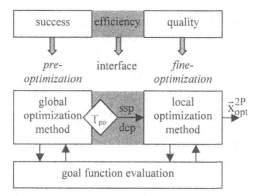

Fig. 3 Basic structure of a combined 2-phase optimization strategy os_{2P}

The result of a combined 2-phase optimization strategy as well as the required computational effort mainly depends on the specific capabilities of the employed global and local optimization method but also on their co-operation. To realize a good co-operation between global and local search, the following problems have to be solved:

- choice of suitable control parameter values for the global optimization method
- choice of an advantageous switch-over point from pre- to fine-optimization (T_{po})
- choice of suitable control parameter values for the local optimization method (dcp)
- choice of a favourable start point $\bar{x}_{start}$ for the local optimization method (ssp)

Our solutions for the problems described above are presented in [10]. Tab. 1 shows some powerful direct optimization methods for global and local search, which are suitable for realization of a combined 2-phase optimization strategy.

	global	**local**
direct optimization methods	Genetic Algorithm GA [3,6] Simulated Annealing SA [1]	Pattern Search PS [5]

Tab. 1 Powerful direct optimization methods for global and local search

Outgoing from the direct optimization methods presented in Tab. 1 there exist the following possibilities to realize a combined 2-phase strategy:

- Genetic Algorithm + Pattern Search (GA+PS)
- Simulated Annealing + Pattern Search (SA+PS)

Both realization alternatives have been already implemented and thoroughly examined in [10].

All optimization strategies considered so far localize only one optimum point when executed. Outgoing from these so-called single-stage optimization strategies, we want to present an optimization algorithm which is able to detect several optimum points of a given multimodal optimization problem. The basic structure of such a multiple-stage optimization algorithm is shown in Fig. 4.

Central component of a multiple-stage optimization strategy os_{ms} is a combined 2-phase strategy os_{2P}. os_{2P} is embedded in an exterior iteration process, which generates step-by-step a sequence of optimum points $\vec{x}_{opt}^1, \vec{x}_{opt}^2, ..., \vec{x}_{opt}^k$, $k \in N$. An iteration step of a multiple-stage optimization strategy is called optimization stage.

os_{ms} stops, if the termination condition T_{ms} is fulfilled. Here, a good termination criterion has been proven to be: stop, if a new or better optimum point could not be located over a specified number of optimization stages. If T_{ms} is not fulfilled, a method called avoidance of reexploration (AR) is applied.

The task of AR is to avoid, that previously found optimum points are located again in subsequent optimization stages. This is done by making already explored regions of the search space unattractive for the global optimization method used for pre-optimization. Therefore, in addition to the goal function values, attractiveness values are introduced and related to each search point of the search space. Attractiveness values are computed by means of an attractiveness function

$$av(\vec{x}) = \prod_{i=1}^{k} [1 - (1 + \alpha \cdot d_i)^{-\beta}], \tag{6}$$

with $d_i = \sqrt{(\vec{x} - \vec{x}_{opt}^i)^2}$;

α, β : scaling factors;

k: number of already found optimum points.

For a further description of multiple-stage optimization and AR we refer to [10].

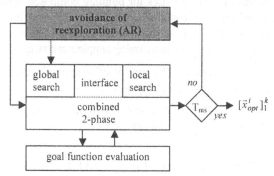

Fig. 4 Basic structure of a multiple-stage optimization strategy os_{ms}

4 Realization

The optimization algorithms presented above have been implemented and integrated into the software tool REMO. REMO running under Solaris on Sun-Sparcstations represents a powerful prototype, which could be successfully applied within many research projects at our institute. This prototype version of REMO can be downloaded at http://goethe.ira.uka.de/~syrjakow/remo.html.

At the moment REMO is completely redesigned in order to make it a user-friendlier tool aimed at professional usage. Our main design goals are the following:

- to extend and further improve the optimization algorithms provided by REMO
- to develop an intuitive graphical user interface in order to ensure an easy usage of REMO for beginners as well as for experts
- to realize a distributed software architecture ensuring accessibility over the World Wide Web and an easy coupling with various modelling tools through a flexible and standardized interface

Especially in order to reach the third goal the former implementation of REMO comprising a Pascal (optimization algorithms) and a C (graphical user interface) part is totally rewritten in Java. Java was chosen because it allows an object-oriented realization, platform-independent execution, Client/Server interoperability via Corba, and accessibility over the World Wide Web via the Java-applet concept. In the following, each of the design goals presented above is described more detailed.

4.1 Extension and Further Improvement of REMO's Optimization Algorithms

As already mentioned the core of REMO is a modular library of powerful direct optimization strategies. These strategies can be divided up into atomic and composed methods. Atomic methods are based on exactly one optimization principle as Genetic Algorithms, Simulated Annealing, and the Pattern Search algorithm. Outgoing from these atomic methods the user can compose more sophisticated algorithms like combined 2-phase or multiple-stage optimization strategies. In our future work we want to extend the existing set of atomic strategies in order to increase the possibilities for building composed optimization methods. Some appropriate and promising atomic strategies would be Evolution Strategies [8] and Tabu Search [2] for global search and new Hill-Climbing methods for local search. Besides adding new atomic optimization strategies we are developing supplementary methods for

- acceleration of the optimization process by goal function approximation [13],
- dealing with goal functions, which are heavily distorted by stochastic noise [4].

These methods should allow a special adaptation of the implemented optimization algorithms to intricate model optimization problems from practice.

4.2   Design of an Intuitive Graphical User Interface

Essential for the user-friendliness of REMO is the design of its graphical user interface (GUI). In the following the most important features of this interface are presented.

When working with REMO the first question usually is which of its many different optimization algorithms is the best one for a given optimization problem. In order to simplify the often difficult choice of an appropriate optimization algorithm we intend to provide the graphical user interface with a wizard, which asks the user some questions about the optimization problem. Based on the given answers the wizard proposes the user one (or more) suitable optimization methods. Such an interview of the user might look as follows: First of all, the wizard will ask, whether there is a priori knowledge about the optimization problem available, or not. Such knowledge about the number of optimum points (only one/several) or characteristic features of the goal function surface (monotonic increase/decrease, etc.) may deliver important hints to choose the right optimization strategy. When we know for example, that there is only a unimodal goal function (goal function with only one optimum point) to optimize, we can instantly exclude all sophisticated global optimization strategies and choose a simple Hill-Climbing algorithm. When we expect a multimodal goal function or do not have any knowledge about the given optimization problem at all, the next question of the wizard would be how many optimum points should be localized. If we are looking for only one optimum point the wizard will propose one of the global optimization methods of REMO. In order to decide which one, the wizard asks us about the optimization quality we are expecting and about the amount of optimization effort we are willing to spend. If we are only interested in a quick and rough approximation of an optimum point the wizard will propose the Genetic Algorithm or Simulated Annealing. In case of an ε-accurate localization a combined 2-phase strategy would be the right choice. When the user wants to find not only one but a sequence of different optimum points the wizard would propose a multiple-stage optimization strategy. In order to come to the final composition of this strategy the wizard would ask us some further questions and so forth.

Summing up, the choice of a suitable optimization strategy depends on the following user information:

- a priori knowledge about the goal function
  - o number of existing optimum points
  - o function surface characteristics
- search space characteristics

- o problem dimension (number of goal function parameters)
- o only discrete parameters
- o only continuous parameters
- o mixture of discrete and continuous parameters
- specified optimization goals
- o improvement compared to a known solution
- o local optimization
- o global optimization
- o multiple-stage optimization
- specified optimization constraints
- o maximum amount of optimization effort

Users having enough expertise to choose a suitable optimization algorithm by themselves can of course deactivate the wizard.

Having chosen an algorithm, the next step is to adapt it to the given optimization problem. This requires a considerable amount of expertise and experience, which usually cannot be expected. For that reason the graphical user interface should provide different user modes for example a beginner, advanced, and an expert mode. In the beginner mode the system makes the necessary decisions mostly by itself. The advanced/expert mode allows experienced users to set many/all control parameters by themselves.

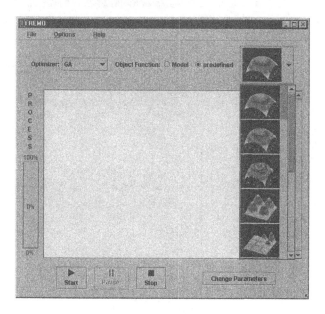

Fig. 5 Graphical user interface of REMO (main window) realized with the Java Swing classes

In the following some important parts of the graphical user interface of REMO are presented. Fig. 5 shows the main window of this interface consisting of the following parts:

- a menu bar comprising the menus "File", "Options", and "Help"
- a combo box to choose an optimization strategy

- two radio buttons to select between the optimization of a model-based goal function and a predefined mathematical goal function (these functions allow the user to make experiments with the different optimization strategies of REMO in order to get familiar with them)
- a progress bar to display the advance of the running optimization process
- a text field to display internal data computed during the running optimization process
- a control part comprising buttons to start (continue), pause, and stop the optimization process
- a button to change the control parameters of the chosen optimization algorithm

The menu "File" enables the user to save/load specified parameter settings and to exit REMO. The menu "Options" allows the user to specify one of three levels of expertise, the seed number for the pseudo random number generator as well as the size of the displayed optimization trace.

By clicking the "Change Parameters" button the window shown in Fig. 6 appears. It allows the user among other things to specify the dimension of the given optimization problem (number of goal function parameters). Beside this, the control parameters of the chosen optimization algorithm (here the Genetic Algorithm) can be adjusted by selecting the tabs "GA-Control", GA-Termination", and "GA-Misc". Here it should be noticed, that the number of adjustable control parameters depends on the chosen level of expertise.

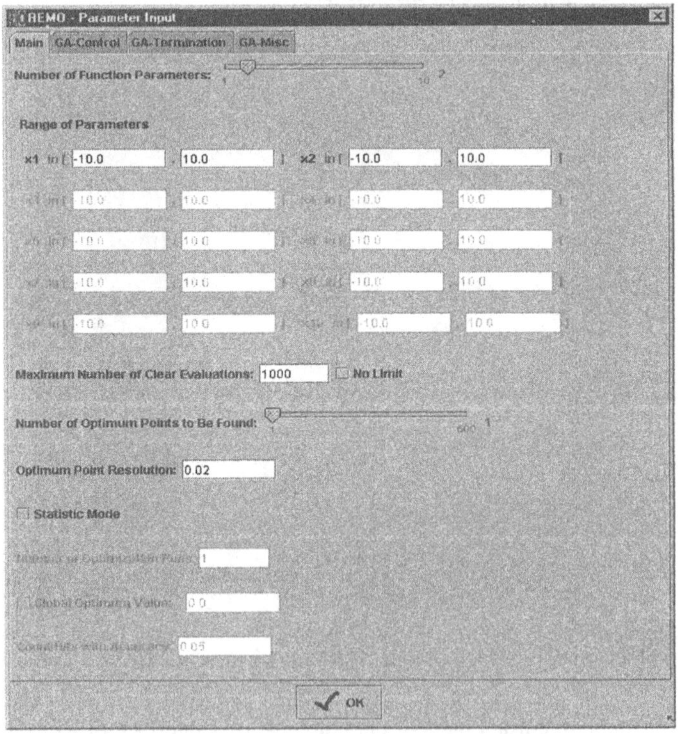

Fig. 6 Window to specify the GA control parameters

4.3   The Distributed Corba-Based Client/Server Architecture of REMO

Functionality for model optimization principally can be either integrated into an existing modelling tool or realized as a stand-alone application. The first alternative is not very promising because of the following reasons:

- the integration of additional functionality into an existing software tool usually is an intricate process resulting in a monolithic application, which is difficult to extend and to maintain
- the optimization functionality is restricted to models of exactly one modelling tool

In order to make REMO a flexible, maintainable, and extensible tool we have decided to realize it as a stand-alone application. Decisive for the flexibility of such an application is an interface, which allows an easy coupling with existing modelling tools, which are required for implementation and evaluation of the models being optimized. For realization of this interface we have decided to use the Common Object Request Broker Architecture (Corba). Corba represents a powerful middleware standard allowing interoperability between computer programs written in different languages like Java, C, C++, etc.

Another design decision was to make REMO available on the WWW. For that reason we have chosen Java, the programming language of the Internet for implementation. The architecture of REMO is presented in Fig. 7. It is a highly distributed architecture, which is based on the business object concept of the OMG (Object Management Group) [7]. Business objects provide a natural way for describing application-independent concepts. A business object is not a monolithic entity. It is factored internally into a set of cooperating objects that can react to different business situations. Business objects are highly flexible and a variation of the Model/View/Controller (MVC) paradigm. They consist of the following three kinds of objects:

- business objects
- business process objects
- one or more presentation objects

Business objects encapsulate the storage, metadata, concurrency, and business rules associated with an active business entity. They also define how the object reacts to changes in the views or model. Business process objects encapsulate the business logic at the enterprise level. In the Corba model, short-lived process functions are handled by the business object. Long-lived processes that involve other business objects are handled by the business process object. Presentation objects represent the object visually to the user. Each business object can have multiple presentations for multiple purposes. The presentations communicate directly with the business object to display data on the screen and sometimes they communicate directly with the process object.

Our developed Client/Server architecture for distributed model optimization consists of the following four tiers [9]:

- Web-client with a REMO presentation object
- Web-server with the REMO business object
- computer A with the REMO business process object
- computer B with a modelling tool

The first tier represents the Web-client with a REMO presentation object, which has the task to display the GUI. This object is connected via the Internet with the REMO business object, which is located on a Web-server in the second tier. In order to keep the workload of the Web-server small, the optimization algorithms of REMO are computed on a third tier where the REMO business process object is located. Finally, on a forth tier the modelling tool is running. The different tiers communicate via Corba ORBs (Object Request Broker).

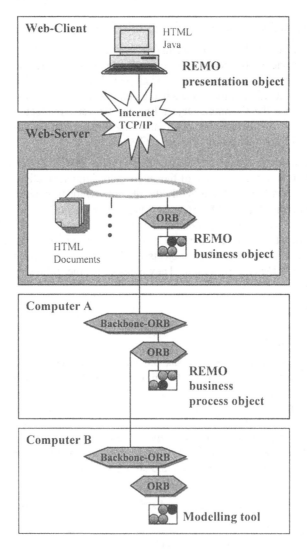

Fig. 7 Overview of the distributed architecture of REMO

5 Conclusions

In this paper efficient and powerful strategies for model optimization were presented. These strategies have been already implemented and integrated into the software tool REMO. REMO could be applied successfully within many projects. Results achieved with this tool are described in [10,11,12, 13].

In the second part of this paper we focused on the ongoing redesign and reimplementation process of REMO. We proposed a highly distributed Client/Server architecture based on Corba, which should allow a flexible coupling of REMO with various modelling tools and its use via the Internet. That way complex simulation models of various kinds can be optimized in a flexible, efficient, and standardized way.

Acknowledgements

We want to thank Prof. D. Schmid for his encouragement and support of our work. We also thank our students, especially H. Renfranz, T. Sommer, D. Haag, J. Gramlich, A. Kehl, and C. Bentz for their engagement and contributions.

Literature

[1] *Aarts, E.; Korst, J.*: Simulated Annealing and Boltzmann Machines: Wiley 1990.

[2] *Glover, F.; Laguna, M.*: Tabu Search: Kluwer Academic Pub. 1997.

[3] *Goldberg, D.E.*: Genetic Algorithms in Search: Optimization and Machine Learning: Addison-Wesley 1989.

[4] *Gramlich, J.*: Optimierung stochastisch verrauschter Zielfunktionen: Diplomarbeit am Institut für Rechnerentwurf und Fehlertoleranz. Universität Karlsruhe 1996.

[5] *Hooke, R.A.; Jeeves, T.A.*: Direct Search Solution for Numerical and Statistical Problems. Journal ACM 8 (1961), pp. 212/221.

[6] *Michalewicz, Z.*: Genetic Algorithms + Data Structures = Evolution Programs: Springer 1992.

[7] *Orfali, R.; Harkey, D.*: Client/Server Programming with JAVA and CORBA: John Wiley & Sons Inc. 1998.

[8] *Schwefel, H.-P.*: Numerical Optimization of Computer Models: Wiley 1981.

[9] *Sommer, T.*: CORBA-basierte Parameteroptimierung von Simulationsmodellen in verteilter Umgebung: Diplomarbeit am Institut für Rechnerentwurf und Fehlertoleranz. Universität Karlsruhe 1999.

[10] *Syrjakow, M.*: Verfahren zur effizienten Parameteroptimierung von Simulationsmodellen: Dissertation am Institut für Rechnerentwurf und Fehlertoleranz der Universität Karlsruhe: Berichte aus der Informatik. Shaker Verlag 1997.

[11] *Syrjakow, M.; Szczerbicka, H.*: REMO - REsearch Model Optimization Package: Tool Descriptions from the 9th International Conference on Modelling Techniques and Tools for Computer Performance Evaluation (Performance Tools'97) and the 7th

International Workshop on Petri Nets and Performance Models (PNPM'97). Saint-Malo, France, June 3-6, 1997, pp. 20/22.

[12]   *Syrjakow, M.; Szczerbicka, H.*: Efficient Parameter Optimization based on Combination of Direct Global and Local Search Methods: in Evolutionary Algorithms (IMA Volumes in Mathematics and Its Applications, Vol. 111), L.D. Davis, K. De Jong, M.D. Vose, L.D. Whitley (eds.), Springer Verlag New York, 1999, pp. 227/249.

[13]   *Syrjakow, M.; Szczerbicka, H.; Berthold, M.R.; Huber, K.-P.*: Acceleration of Direct Model Optimization Methods by Function Approximation: Proceedings of the 8th European Simulation Symposium (ESS'96). Genoa, Italy, October 24-26, 1996, Volume II, pp. 181/186.

Quantitative Measures for Systematic Optimization, Validation, and Imperfection Compensation in the Holistic Modeling and Parsimonious Design of Application-Specific Vision and Cognition Systems

Doz. Dr.-Ing. Andreas König, TU Dresden, Germany
Dipl.Wirtsch.-Ing. Michael Eberhardt, TU Dresden, Germany
Dipl.-Ing. Jens Döge, TU Dresden, Germany
Dipl.-Ing. Jan Skribanowitz, TU Dresden, Germany
Andre Kröhnert, TU Dresden, Germany
Andre Günther, TU Dresden, Germany
Robert Wenzel, TU Dresden, Germany
Tilo Grohmann, TU Dresden, Germany

Abstract

Intelligent systems, e.g., for vision and cognition tasks, enjoy increasing industrial acceptance and application. Application domains range from optical character and handwriting recognition to biometric systems. The joint exploitation of advanced microelectronics, sensor technology, and intelligent systems provides a tremendous economic potential. Tight application constraints such as, e.g., size, speed, performance, and power consumption give increasing attraction to dedicated integrated system implementations exploiting bio-inspiration and opportunistic design techniques in analog or mixed-signal circuits and systems. However, to achieve a viable design at reasonable cost and time-to-market, an appropriate design methodology is required. This paper presents quantitative measures for system-oriented imperfection compensation, optimization, and validation of dedicated application-specific intelligent systems. These measures serve for the fast and consistent behavioral modeling of an aspired intelligent system, and support the rapid and consistent transformation into a physical design. Further, they provide the basis for ensuing design automation of the process, and contribute to the ongoing vivid activities of method and tool development for system simulation and evaluation.

1 Introduction

Intelligent systems, e.g., for vision and cognition tasks, enjoy increasing industrial acceptance and application. To name only a few examples, character and handwriting recognition, industrial visual inspection and identification, surveillance and identification, access control using biometric systems as well as object detection and tracking tasks, e.g., eye or face tracking, passenger counting, aircraft docking, or driver assistance tasks in the automotive domain, are subject of industrial exploitation efforts. A plethora of economically attractive tasks can be solved today by available general-purpose (GP) hardware. The range of applications amenable to GP hardware solutions will continue to grow, due to the ongoing validity of Moore's law and the vivid design activities driven by, e.g., the surging communication market and respective low-voltage and low-power implementations. However, numerous machine vision problems, for example complex surveillance tasks, automotive applications, multi-media interactive applications, biometric tasks, advanced image coding, or automated in-line visual inspection and vis-

R. Merker and W. Schwarz (eds.), System Design Automation, 247-258.
© 2001 *Kluwer Academic Publishers.*

ual process control, impose high demands on viable solutions in terms of size, speed, data-throughput, performance, and power consumption. Furthermore, development cost and time-to-market are critical factors. Today's predominantly digital systems, e.g., DSP arrays or dedicated implementations, as BGT's *Systolic-Array-Processor* (SAP) with a strong military background, cannot always provide an adequate problem solution by available state-of-the-art hardware with regard to all constraints specified above.

As generic systems are still out of the question today, dedicated integrated hybrid system implementations exploiting bio-inspiration as well as an opportunistic and parsimonious design style, especially incorporating CMOS compatible sensor technology, offer a viable alternative for application-specific embedded solutions [1]. Thus, this design approach, pioneered by Eric Vittoz [1], attracted researchers and led to a large number of design activities. Most of these, e.g., from the domain of vision-chips, were bottom-up design efforts conducted by people with strong technological background. Only a few activities featured a system-oriented view and led to viable implementations of commercial interest. Rare examples are, e.g., CSEM's motion detector chip for pointing devices of Vittoz et al., that is used in Logitech's Marble trackballs [1], or Mitsubishi's "artificial retina" chip and related systems for 3D human motion recognition, interactive games as well as general image preprocessing tasks [2].

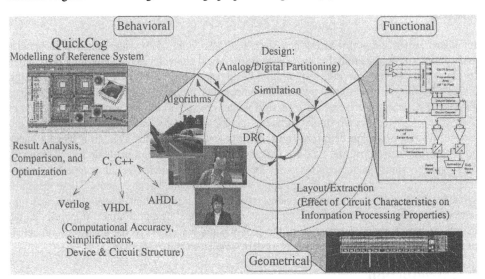

Fig. 1 Concept of the design methodology illustrated by an extended Y-diagram.

Learning from these activities as well as the surge of neurocomputer implementations and the outcome of these activities in the beginning of the last decade (see, e.g., [3]), a top-down design methodology was developed in our work in the last years [4]. Our methodology, illustrated in **Fig. 1** by an extension of the well-known Y-diagram, pursues a system-oriented design style and enhances the standard design flow by a level for fast behavioral modeling of the vision and cognition task, denoted as **QuickCog** [4] [5]. Inherent quality measures and optimization mechanisms in our approach accelerate and partly automate the design. Also, they support the designer to meet constraints of turnaround time, time-to-market, development and design cost and to ensure overall system validity and performance. **Figure 2** illustrates the currently implemented design activities and interfaces to Cadence DFW II. Our behavioral modeling tool for cognitive systems is linked with the standard design hierarchy of integrated circuits and systems. As a dedicated instance of an IC design tool we choose Cadence DFW II. One or several blocks of the behavioral model in **QuickCog** are designed within DFW II. Stimuli as well as parameters generated in system training, e.g., neural network weights, are transferred to DFW II tools, e.g., as behavioral cell views in Verilog-A. Simulations can take place on circuit level, based on both the circuit schematic view and the extracted view from the layout. Result data is exported and converted via a second interface so that QuickCog can import the result data, evaluate it, and compare it with the results achieved by the behavioral reference system. Thus, circuits and systems can systematically be evaluated, accepted or discarded for potential implementation, and also be optimized with regard to the desired performance. In **Fig. 2**, the basic flow is illustrated, that will be employed in the practical experiments reported in this paper.

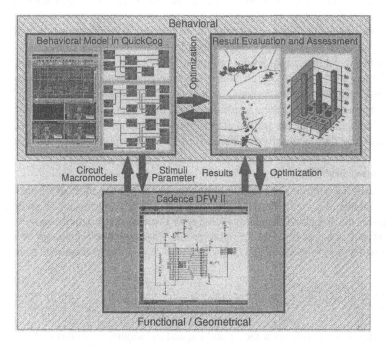

Fig. 2 Interface and data flow between **QuickCog** and Cadence DFW II.

The key issue for a successful implementation of an optimization concept is to find the appropriate cost functions. For instance, in image acquisition and processing, there is a definite bifurcation between the consumer market and the machine vision market with regard to salient characteristics and quantitative assessment. In the first case, quality demands imposed by human visual perception must be met, and in the latter case, the overall system discriminance and reliability must be optimized. Here, we focus on the latter case, and discuss related quantitative measures in the next section. Their application to optimized circuit and system design will be elucidated using examples from our work on vision and cognition systems. Implications for applicable optimization strategies and system design automation as well as methods and tools for system simulation and evaluation will be pointed out. Also, issues of compensating hardware imperfections and nonidealities by adaptation or learning in a single or over several system slabs, similar to the well-known *chip-in-the-loop-learning* (CHILL) concept [6], will be addressed. The export of extracted circuit macro models for inclusion in the learning process to achieve compensation of circuit imperfections is indicated in **Fig. 2**.

2 Quantitative Measures

In this section, we will describe and discuss suitable measures for system discriminance and reliability assessment. Special emphasis is on measures, that take into account the multivariate statistical nature of the information representation in vision and cognition systems. Though image related comparison measures, e.g., difference images or signal-to-noise-ratios (SNR), can be salient, and give some idea of imposed distortions by a certain hardware implementation, e.g., for image filtering or flow-field computation, no reliable information on the system-level effect can be extracted in general. More reliable and statistically meaningful information can be extracted based on a whole set of representative images or feature vectors, denoted as a sample set in the following. For a given relevant sample set, salient information with regard to, e.g., signal preservation, topology preservation, distance preservation, or discriminance preservation can be gained for system-level assessment, both from feature space visualization and apt numerical measures. Such measures, which will briefly be treated in the following, can serve both for fast reference system design as well as continued comparison between reference system and ongoing hardware system implementation. Both overall thresholded system performance, e.g., recognition rate, as well as graceful performance degradation due to chosen circuit imperfections can thus be observed and employed for design optimization and automation.

2.1 Unsupervised Measures

Measures working without supervised information just assess the effect of a transformation on the data contents, e.g., the distortion introduced by image compression or coding. As mentioned above, the computed degree of topology or distance preservation can serve for system assessment and as a cost function for optimization. Distance preservation can be measured by assessing patterns in m-space $\mathbf{X}$ and d-space $\mathbf{Y}$ with regard to the change of their mutual distance:

$$E = \frac{1}{c} \sum_{i=1}^{N} \sum_{j=i+1}^{N} \frac{(d_{X_{ij}} - d_{Y_{ij}})^2}{d_{X_{ij}}} ; \quad c = \sum_{i=1}^{N} \sum_{j=i+1}^{N} d_{x_{ij}} ; \quad d_{X_{ij}} = \sqrt{\sum_{q=1}^{m} (x_{iq} - x_{jq})^2} ; \quad d_{Y_{ij}} = \sqrt{\sum_{q=1}^{d} (y_{iq} - y_{jq})^2} .$$

Here, c denotes a normalization factor and d_{Xij} and d_{Yij} denote the distances between pattern_i and pattern_j, in space $\mathbf{X}$ and $\mathbf{Y}$, respectively. This measure corresponds to the cost function used in Sammon's Non-Linear-Mapping (NLM), and, for $d = m$, gives a measure of the distortion or preservation of the data structure due to constrained implementation.

In addition, topology preservation can be measured, for instance by comparing a number k of nearest neighbor patterns in m-space $\mathbf{X}$ and d-space $\mathbf{Y}$ with regard to the coincidence of the nearest neighbor rank NN_{Xij} and NN_{Yij} in $\mathbf{X}$ and $\mathbf{Y}$ space, respectively. For each coincidence of rank positions, a local measure for each pattern q_{mj}^* is incremented. A global topology measure q_m^* is given by

$$q_m^* = \frac{1}{kN} \sum_{j=1}^{N} q_{mj}^* .$$

Variations of this measure can be introduced, which, e.g., respond in proportion to deviations in the rank position by graceful degradation of the q_m measure [10]. Both described measures allow the assessment of data or sample sets, e.g., by comparing ideal simulation results with those of a constrained hardware realization for $d = m$ with regard to data structure preservation.

2.2 Supervised Measures

In pattern recognition problems, supervised information for the decision problem, i.e., class affiliation, complements the image or feature data. In contrast to the previous section, perturbations of the feature data can be tolerated if the decision making capability of the system is not affected. This means, that, instead of structure preservation, discriminance preservation is now in the focus of interest. Numerous measures can assess the discriminance of a recognition system. The most straightforward measure is the classification or recognition rate $\mathbf{R}$ provided by a system. The decrease in recognition rate observed, e.g., by comparing ideal simulation results with those of a constrained hardware realization, can serve as a metric for the regarded system, its expected performance, and the consistency of the current design. However, $\mathbf{R}$ is giving no idea how close the classifier implementation was to failure. As the classifier decision is computed by a maximum operation on the class-specific a posteriori probabilities, a deterioration with regard to the dissimilarity of the a posteriori probabilities will not be visible. Thus, a contrast metric q_k was introduced as follows:

$$q_{ki} = \frac{1}{N} \sum_{i=1}^{N} |O_{C1} - O_{C2}|; \qquad O_{C1} = \max_{j=1}^{L} O_j; \qquad O_{C2} = \max_{j=1;C2 \neq C1}^{L} O_j.$$

This measure discloses information of the implementation effect on the certainty and reliability of the classification. As a wrong classification with a high contrast is, of course, detrimental, an improved contrast measure q_k' inverts the sign of the partial sums $|O_{C1} - O_{C2}|$ for misclassifications. Also, the visualization techniques discussed in Subsection 2.3 can be salient for analysis and optimization (see **Fig. 6** (right)).

In addition to the overall recognition rate $\mathbf{R}$, the class specific errors due to confusion between classes can be evaluated by confusion matrices [5], gaining clues for the assessment and comparison of different system implementations. Regions in feature space with the underlying feature value combinations that are most susceptible to misclassification are disclosed. Interesting conclusions can be drawn, e.g., on the chosen circuits' susceptibility with regard to feature value ranges. For incremental assessment of the individual processing stages in an intelligent system, only the recognition rate $\mathbf{R}$ does not suffice. To assess feature computation methods, direct assessment capability of the resulting feature space with regard to discriminance is required. The intercluster and the intracluster distances as well as class region's overlap, compactness, or separability are salient characteristics to quantify discriminance and especially changes in discriminance. In this work robust, nonparametric nearest-neighbor techniques based on the Reduced-Nearest-Neighbor-Classifier (RNN) and the Edited-Nearest-Neighbor-Classifier (ENN) are employed [5]. A separability measure q_{si} is computed by

$$q_{si} = \frac{1}{L} \sum_{i=1}^{L} \frac{N_i - (T_{RNN_i} - 1)}{N_i}$$

for an L-class problem. Here, N_i denotes the number of patterns affiliated to class ω_i, and T_{RNNi} denotes the number of reference patterns selected by the RNN method with class affiliation ω_i. The measure q_{si} returns 1.0 for linear separability, and degrades towards zero with increasing complexity of the class border. This measure is very fast due to its O(N) complexity, but assumes a rather coarse value range. To detect gradual changes in the feature set, the overlap measure q_o is computed by examining a fixed neighborhood of k neighbors for each point. In the most simple case q_o is computed from the ratio of neighbors k^* with the same class affiliation and the number k of all regarded neighbors for each pattern by

$$q_o = \frac{1}{L}\sum_{i=1}^{L}\frac{1}{N_i}\sum_{j=1}^{N_i}\frac{k^*}{k}.$$

An even more finely grained measure can be achieved by including the distance information between neighbors in the ratio computation [5]. Even slight expansions and contractions in feature space can then be detected and quantified [5]. Again, q_o returns 1.0, indicating no overlap and degrades towards zero with increasing overlap in pattern space. Summarizing, the measures q_{si} and q_o provide metrics to probe and assess discriminance at an arbitrary point in the processing chain of a recognition system. These metrics can be employed to optimize a system under design, e.g., by choosing the best method for feature computation, to select the most salient features, and to optimize method parameters [5]. Vision and cognition system design automation, employing such cost functions and optimization schemes, e.g., genetic algorithms and evolutionary strategies, to speed up the application-specific system development, is a research goal currently pursued in our work [5]. After thus achieving a behavioral implementation of the desired system, the same assessment scheme can be employed for the physical design of an integrated system solution. This will be pointed out in Section 3.

2.3 Feature Space Visualization

In addition to the described quantitative metrics, advanced information visualization techniques can be employed to display, analyze, and especially compare arbitrary features spaces. **Fig. 3** (left) gives the basic idea of the feature space visualization concept. By a dimensionality reduc-

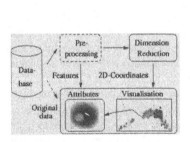
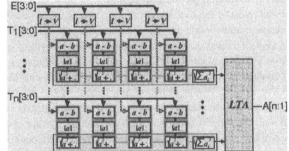

Fig. 3 Concept of feature space visualization (left), RNN recall classifier structure (right).

ing mapping [10][5], which is out of the scope of this paper to discuss here, it is feasible to obtain a two-dimensional representation of the m-dimensional feature space. Each projection point in the display corresponds to a pattern in feature space. Class affiliation is expressed by a color or text attribute. Thus, clusters as well as intracluster and intercluster distances, overlap, and separability can be observed in the visualization. Further, misclassifications can be marked in feature space, so that, in addition to the statistical rate information, the location of misclassified patterns in feature space becomes overt. This allows to analyze context information, e.g., the pattern neighborhood or the underlying feature value combination for this pattern. This makes it possible to draw interesting conclusions on, e.g., a certain circuit's performance, and supports trouble shooting in the design process, as will be demonstrated in the next section.

3 Application Examples

3.1 Assessment of Classifier Performance

As the first example of the introduced metrics and evaluation schemes, a performance assessment of a classifier implemented as a reference software system and its corresponding schematic and layout circuit implementation will be carried out. For demonstration purposes, we resort to a simple problem, denoted as Iris data, which has become a widespread and well accepted benchmark problem for classification methods. The data was collected from Iris flowers to distinguish three species setosa, virginica, and versicolor by means of the four features given by setal and petal length and width. The approach can straightforward be generalized to real-world applications, e.g., facial feature classification from image sequences for an eye-tracker implementation [8].

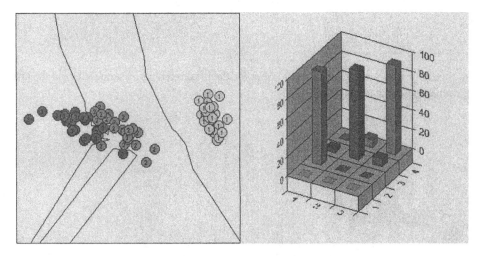

Fig. 4 Visualization of Iris training data set (left), reference system results for Iris test data (right).

In this example, we use the RNN classifier, given for recall in **Fig. 3** (right). In contrast to the kNN, the RNN selects a subset of salient prototype vectors from the training data set in an iterative training or learning phase. This training process is carried out in the **QuickCog** system, which serves for reference system modelling. For the Iris data training set with 75 sample vectors, given in **Fig. 4** (left), 10 prototypes are chosen by the RNN training algorithm. A test set of another 75 vectors is now applied to assess the generalization capability of the RNN classifier in the ideal reference system. A recognition rate **R** of 93.33% was achieved by the software classifier implementation. The confusion matrix in **Fig. 4** (right) shows that one pattern of class 1 was classified as class 2, two patterns of class 2 were classified as class 3, and two patterns of class 3 were classified as class 2. In excess of this information, **Fig. 5** shows the pattern positions in feature space tagged by exclamation marks.

The feature space visualization tool allows rapidly to identify the patterns 61, 51, 50, 14, and 30 as the ones having been misclassified by the reference system implementation. Following the outline of our design methodology, given in **Fig. 1,** the RNN recall classifier was designed as an analog circuit. First, advancing from behavioral to functional representation, the classifier structure, given in **Fig. 3** (right), was designed on schematic level and simulated.

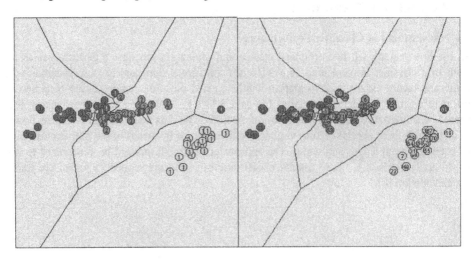

Fig. 5 Feature space with classification errors for the ideal reference system (left) and the analog circuit implementation (right).

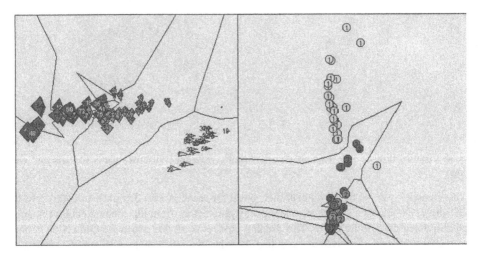

Fig. 6 Feature space with classification errors of first-cut classifier design and feature value plots (left), visualization by a posteriori probabilities (right).

After some refinements of the device dimensions, $R = 92\%$ could be achieved. The feature space visualization, given in **Fig. 5** (right), discloses the identity of the additionally misclassified patterns. Now patterns 61, 51, 45, 42, 30, and 14 are misclassified, i.e., pattern 50 is now correctly classified whereas 42 and 45 are misclassified due to the influence of the implementation. A tentative layout section and complete RNN layout implemented using AMS Int. AG's 0.6 μm CMOS technology is given in **Fig. 7**. One feature slice consumes $77.4 \times 22.2 \, \mu m^2$ and

10 nW of power at a rate of 1k patterns/sec. The modular layout of the cells allows both to cascade the computational blocks vertically and horizontally to achieve higher feature or weight vector numbers as well as to achieve larger transistors and higher accuracy, respectively. A modular cell library for classifiers, e.g., BP, RCE, RBF, PNN, or kNN, as well as feature extraction for low-power recognition system design is currently advanced in our work [9].

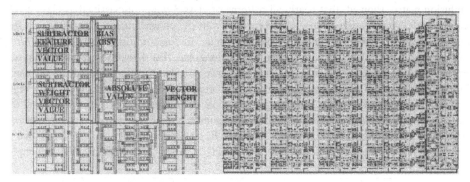

Fig. 7 Layout section of RNN implementation (left), complete Iris-RNN classifier layout (right).

The deviations that were disclosed by the investigation and those additional effects imposed by layout design will be dealt with using appropriate circuit models in the learning phase (see **Fig. 2** and Section 4) in upcoming work. Deviations due to process parameter variations in manufacturing must be assessed in the same way and dealt with CHILL to achieve a consistent yet parsimonious design. In **Fig. 8**, a complete layout of an RNN-classifier for eye-shape recognition is shown, that uses the Euclidean distance, six features, and five reference vectors. This classifier was designed to serve in a low-power eye-tracker [8]. Features are computed by selected Gabor filters from gray-value image blocks containing eye-shapes and non-eye-shapes.

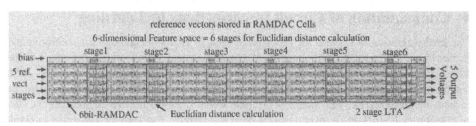

Fig. 8 Complete layout of RNN-classifier for eye-shape recognition in an eye-tracker system.

For weight storage, 6 Bit RAMDACs were employed, which consumed approximately 80% of the area for the Manhattan distance version. In [9] it was shown, that the more convenient and simpler to implement Manhattan distance could be employed. However, one additional reference vector was required, when RNN learning was carried out with this model constraint. Both implementations have a rather low speed of 1k patterns/s. The Euclidean distance version has the size of 1595 µm x 184 µm and consumes an average of 348.7 nW/pattern. The Manhattan distance version has the size of 1460 µm x 215 µm and consumes an average of 277 nW/pattern. The latter version achieved **R** = 100% for training and test set, requiring no further optimization in the design phase. Individual manufactured samples could also be subject to CHILL for a performance boost.

3.2   Classifier Error Visualization and Certainty Assessment

In an earlier phase of the work, a first-cut design only returned $\mathbf{R}=73\%$. An analysis of the confusion matrix showed, that nearly all patterns of class 3 were misclassified as class 2. The feature space visualization of the classifier error, with the enhancement of displaying all four feature values in a radial plot at each projection point, discloses in **Fig. 6** (left), that the errors dominantly occur for large feature values. This gives a clue, that the chosen circuit cannot deal properly with large feature values or large differences of feature values. Based on this cue, a flaw in the subtractor circuit was detected and corrected, and the results already given in **Fig. 5** (right) where obtained. This is a very simple example of how quantitative assessment and feature space visualization can assist in vision and cognition circuit and system validation and optimization. In addition to the assessment based on $\mathbf{R}$, classifier certainty can give additional information how close the classifier was to failure using q_k and q_k', even if $\mathbf{R}$ is still satisfying.

Reference system	Circuit schematic	Circuit layout
$q_k = 0.54268$	$q_k = 0.29393$	$q_k = 0.29366$
$q_{k'} = 0.53256$	$q_{k'} = 0.28939$	$q_{k'} = 0.28912$

Table 1 Assessment of a posteriori probabilities for reference, schematic, and extracted layout.

This is especially of importance with regard to manufacturing effects on circuits, as small safety margins in the classifier certainty can lead to misclassifications in the manufactured circuit. In **Table 1** the results of the quantitative assessment of the reference system and the simulated circuit implementation are shown. Though, $\mathbf{R}=92\%$ is achieved again, an increasing contrast degradation from reference over schematic to layout due to the effects of the chosen implementation can be observed. A-posteriori values can be visualized as given in **Fig. 6** (right).

4   Compensation of Circuit Imperfections by Learning

The implementation of systems for recognition tasks in analog technology suffers from imperfections imposed by the chosen circuit itself as well as by the manufacturing process. Learning schemes can serve for the compensation of these detrimental effects. Neural networks for instance are famous for their fault tolerance and the property of graceful degradation. Also, due to inherent plasticity, i.e., the change of synaptic weights, adaptation can compensate the effect of a lesion or a stuck-at-error in terms of microelectronics. The learning algorithm integrates the imperfectly working circuit and tries to cope with the problem by appropriate adjustment of the learning parameters. Intel employed this concept for their ETANN (Electrically Trainable Analog Neural Network) chip, that implements 64 analog neurons with approximately 2×64 synapses per neuron for Hopfield or multi-layer feedforward network computation. Weight storage takes place within the Gilbert multipliers, employing EEPROM technology for analog weight storage [6]. Achievable synaptic accuracy and thus recognition results can be improved by *chip-in-the-loop-learning* [6] (CHILL), that employs the chip and its current weight values in the forward-phase of backpropagation learning. As sketched in **Fig. 9**, weight updates are computed based on the information of the individual chip's device configuration and flaws, and are used to correct on-chip weights. This leads to a significant compensation of the individual chip's imperfections and an increase in performance [6]. This concept can be generalized and employed in the design phase of vision and cognition systems [9]. The constraints of regarded individual circuits, e.g., a simplification of the distance metric using Manhattan instead of Euclidean distance, can be incorporated in the learning algorithm. The learning algorithm can now compensate the imperfections and constraints by, e.g., appropriately modifying the classifier topology and parameters. It is a well known property of dynamically growing network topologies, e.g., cascade correlation, RCE, or RNN, that an increasing number of hidden neurons is included in the case of constrained accuracy or metric. In addition to classifier topology and parameters, e.g., weights, also VLSI design parameters could be subject to the optimization process. Information of circuit properties with regard to imperfections and nonidealities will be transferred to the learning system using appropriate macro models, which makes macro modeling [7] in this context an issue of interest in our ongoing research. Thus, circuits with salient properties, e.g., with regard to power dissipation as well as area consumption, but limited performance due to imperfections, can be tailored to the problem with this optimization approach, while retaining performance of the overall system in terms of the chosen metrics. The proposed collective learning and optimization approach can be applied to one or several system slabs, and is not constrained to derivable cost functions for gradient descent optimization. For instance, the cost functions introduced in Section 2 can serve with techniques from evolutionary computing for improved system design, e.g., from feature computation to classification. **Fig. 9** gives an example of parameter optimization in feature computation employing q_s and q_o [5].

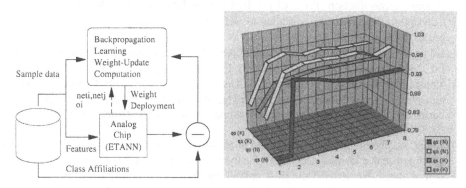

Fig. 9 Concept of chip-in-the-loop-learning (left) and general parameter optimization (right).

5 Conclusions

In this work, we have discussed quantitative measures for assessment of intermediate and final processing results in vision and cognition systems. These metrics can be employed for systematic and fast design and optimization of application-specific system solutions [5]. The approach can be generalized to the fast and consistent physical design of the regarded systems. Linking with the standard design hierarchy, the introduced metrics and the design platform **QuickCog** serve for validation and optimization purposes in the parsimonious design of application-specific systems. Salient novel circuits and opportunistic design practice can be fully exploited by our approach. A first successful demonstration using Iris benchmark data as well as application data from the problem of eye-tracking was given in this paper. In ongoing work, the methodology is applied and extended using the design of a low-power eye-tracker system as a research vehicle. A combination of our design methodology with general activities for analog and mixed-signal design automation is considered as highly desirable. Considerable challenges in addition to manufacturing deviations are given by dynamic deviations, e.g., due to temperature drift. CHILL alone cannot cope with these, and the investigation of bio-inspired robust signal representations and circuits as well as on-line adaptation mechanisms will thus be a focus of our future work.

6 Acknowledgments

The reported work has been carried out in the project *Holistic Modeling of Integrated Low-Power Recognition Systems* (GAME, GAnzheitliche Modellierung integrierter verlustarmer Erkennungssysteme) which is funded within the scope of the research program VIVA (Grundlagen und Verfahren verlustarmer Informationsverarbeitung, SPP 1076) by the German research foundation "Deutsche Forschungsgemeinschaft". All responsibility for this paper is with the authors.

7 Literature

[1] Vittoz, E.: Present and Future Industrial Application of Bio-Inspired VLSI-Systems. Proc. of 7th Int. Conf. On Microelectronics for Neural, Fuzzy, and Bio-Inspired Systems, Granada, ISBN 0-7695-0043-9, IEEE CS, pp. 2 -11, 1999.

[2] Kage, H., Funatsu, E., Tanaka, K., Kyuma, K..: Artificial Retina Chips for 3-D Human Motion Reconstruction. Proc. of the IIZUKA'98 Int. Conf. On Methodologies for the Conception, Design, and Application of Soft-Computing. Iizuka, Japan, World Scientific, pp. 76, 1998.

[3] König, A..: Survey and Current Status of Neural Network Hardware. Invited. Proc. of 5th Int. Conf. On Artificial Neural ICANN'95, Paris, France, pp. 391 - 410, 1995.

[4] König, A, Skribanowitz, J., Schreiter, J., Getzlaff, S., Eberhardt, M., Wenzel, R.: Ein System zur schnellen Modellierung von Bildverarbeitungs- und Erkennungssystemen. Tagungsband Dresdner Arbeitstagung Schaltungs- und Systementwurf DASS'98, FhG IIS-EAS, 1998.

[5] König, A, Eberhardt, M., Wenzel, R.: QuickCog Self-Learning Recognition System – Exploiting machine learning techniques for transparent and fast industrial recognition system design. Image Processing Europe, PennWell, pp. 10 – 19, Sept/Oct. 1999.

[6] Intel Corp.: Intel 80170NX Electrically Trainable Analog Neural Network. Product Description and Data Sheet, Intel Corp. Order No. 290408-002, 38 pp, June 1991.

[7] Einwich, K., Haase, J., Prescher, R., Schwarz, P.: Makromodellierung für Mixed-Signal-Schaltungen. mikroelektronik me Bd.9 H.4, VDE-Verlag Berlin, pp. 20-25, 1995.

[8] König, A., Günther, A., Kröhnert, A., Grohmann, T., Döge, J., Eberhardt, M.: Holistic Modelling and Parsimonious Design of Low-Power Integrated Vision and Cognition Systems. Proc. of the IIZUKA'00 6th Int. Conf. On Methodologies for the Conception, Design, and Application of Soft-Computing. Iizuka, Japan, World Scientific, pp. , 2000.

[9] König, A., Günther, A., Döge, J., Eberhardt, M.: A Cell Library of Scalable Neural Network Classifier for Rapid Low-Power Vision and Cognition Systems Design. Proc. of 4th Int. Conf. On Knowledge-Based Intelligent Engineering Systems & Allied Technologies KES2000. Brighton, UK, pp. 275-282, 2000.

[10] König, A.: Interactive Visualization and Analysis of Hierarchical Neural Projections for Data Mining. IEEE TNN, Vol.11, No.3, Special Issue on Neural Networks for Data Mining and Knowledge Discovery, pp. 615-624, 2000.

Index

Erratum

System Design Automation
Edited by R. Merker and W. Schwarz
ISBN 0-7923-7313-8

Please find here the missing pages of the article by U. Hatnik et al.,
p. 185, following page 194.